A New GeoComputation Pattern
and Its Application in Dual-Evaluation

Wensheng Zhou

A New GeoComputation Pattern and Its Application in Dual-Evaluation

Surveying and Mapping Press

Springer

Wensheng Zhou
School of Architecture
Tsinghua University
Beijing, China

Supported by the National Key Research and Development Plan (2018YFD11003)

ISBN 978-981-33-6434-9   ISBN 978-981-33-6432-5   (eBook)
https://doi.org/10.1007/978-981-33-6432-5

Jointly published with Surveying and Mapping Press
The print edition is not for sale in China (Mainland). Customers from China (Mainland) please order the print book from: Surveying and Mapping Press.

This Springer imprint is published by the registered company Springer Nature Singapore Pte Ltd.
The registered company address is: 152 Beach Road, #21-01/04 Gateway East, Singapore 189721, Singapore

Sharp tools make good work.
工欲善其事, 必先利其器

Foreword

Since the 1960s, geographic information system (GIS) technology began to rise and gradually has been widely used in China, with remarkable application results. In spite of the fact that the application of GIS in visual management and public service is relatively mature, and there are two deficiencies in business application. First, for large-scale GeoComputation (GC) tasks involving a wide range and many contents, there is no way to verify the quality of their GC results, and only comparative computation can be used to verify each other. Second, in scientific research institutions or enterprises and institutions, general technicians or researchers cannot easily construct, maintain, and use the spatial processing model by themselves. The root cause of these problems lies in the fact that although GIS has developed for more than 50 years it still has not solved the problem of easy use of GIS.

On this issue, Prof. Wensheng Zhou of Tsinghua University has made pioneering work. In the process of compiling the *Technical Guidelines for Evaluation of Carrying Capacity of Resources and Environment and Suitability for Territorial and Spatial Development*, aiming at the characteristics of Dual-evaluation computation, after systematically analyzing various problems existing in the current GC pattern, a brand-new GC pattern was proposed to solve the above two problems. Since the new GC pattern is a transparent and table-based spatial data processing pattern, we can trace back each processing step of the whole data procession. Therefore, GC can ensure the correctness of the processing results under the condition that the input data is accurate and the data processing flow is correct, and the computation results can be repeated. Secondly, because the GC language close to natural language is used to describe the whole process of spatial data processing in MS Word environment, the difficulty of spatial modeling is greatly reduced, and ordinary technicians or researchers without programming experience can build, maintain, and use spatial data processing models according to their own business requirements. This new GC pattern makes GIS from traditional spatial data visualization tool to real and practical spatial data analysis tool and makes GIS "aristocratic" application to "civilian" application, which plays an important role in promoting the development of GIS.

As this new GC pattern has passed the test of dual-evaluation work, its effectiveness and practicability have been proved. It is hoped that this pattern can be widely popularized and applied in GIS application field.

The author has been exploring the work of spatial information technology in urban and rural planning and cultural heritage protection all the year round and has achieved remarkable research results, opening a new field of GIS application. He hopes that more scholars or technicians in related fields will join in it and make contributions to the development of GIS in China and even in the world.

April 2019

Jiancheng Li
Academician of Chinese Academy of Engineering
Wuhan, China

Preface

At present, the "Evaluation of the Carrying Capacity of Resources and Environment and the Suitability of territorial and spatial Development" (hereinafter referred to as dual-evaluation) is a basic work serving the compilation of territorial and spatial planning. Because the dual-evaluation work has the characteristics of wide implementation scope (including the work at the national, provincial, city, and county levels), involving many specialties (including ecological, land resources, water resources, environment, climate, disasters, location, and other evaluation contents), and difficult data processing (nearly 400 GC processes are required). How to carry out this work scientifically, systematically, efficiently, and standardized under the guidance of the *Technical Guidelines for Evaluation of Carrying Capacity of Resources and Environment and Suitability for Territorial and Spatial Development* (hereinafter referred to as *Technical Guidelines*), so as to make the quality of GC results controllable, inspectable and traceable, and the GC work easy to implement, has become an important topic of concern in the dual-evaluation work.

However, the traditional GC pattern, whether toolbox pattern, visual modeling pattern, script development pattern, or independent system development pattern, cannot meet the requirements of dual-evaluation work.

Therefore, under the inspiration of end-user programming thought, this paper presents a new GC pattern called Document As A System (DAS). The core of this computation pattern is a set of self-created GC language (G language for short). The language describes the GC process in a standardized way in the form of natural language, allowing technicians to conveniently control the GC process and parameters through the GC language in technical documents, thus realizing full automation of GC.

In the traditional computation pattern, the technical documents only guide the technicians how to operate, and the GC results depend entirely on the level and sense of responsibility of the technicians. However, DAS pattern intelligentizes the technical documents with G language, and the G language interpreter directly reads the contents of the technical document, thus accurately executing the specified GC. The whole GC process is not only highly automated, but also standardized and transparent, thus realizing the goals of high efficiency, standardization, traceability, and easy implementation of GC.

Compared with the traditional computation pattern, DAS pattern has the important technical characteristics of transparent computation, controllable results, traceability, and easy implementation, which is a change to the existing GC pattern. In DAS pattern, geographical analysis models with different degrees of complexity can be constructed by using G language, Moreover, due to the direct correspondence between the limited key statements of GC and natural language, technicians can build geographic analysis models in their familiar MS Word or Kingsoft WPS documents at will to solve geographic analysis problems in daily business or research work under the condition of almost no learning cost.

This book systematically introduces the relevant knowledge of DAS pattern and takes the dual-evaluation of territorial and spatial planning as an example. According to the *Technical Guidelines*, the GC process of evaluation content in the *Technical Guidelines* is completely and standardized described by G language. Specifically, the book includes three parts: theory, practice, and dual-evaluation GC models. The first part is the theoretical part, which focuses on combing the research related to territorial spatial planning, dual-evaluation of territorial and spatial planning, GC and programming language. On this basis, it introduces DAS pattern and G language-related theories. The second part is the practical part, which focuses on the interpreter of G language, the integrated development environment of G language, the compiling and debugging of intelligent documents, and the development, installation, and use instructions of the dual-evaluation system. The last part is the GC model; it introduces the dual-evaluation GC models described by G language; it mainly includes individual evaluation (ecological evaluation, land resources evaluation, water resources evaluation, climate evaluation, environmental evaluation, disaster evaluation, and comprehensive advantage evaluation), Integrated evaluation (importance of ecological protection, suitability of agricultural production, suitability of urban construction), optional evaluation (suitability of marine development and utilization, importance of cultural protection, and suitability of mineral resources development), comparative analysis (current situation comparison, three-lines comparison, provincial and municipal comparison, and other comparisons), map analysis (thematic atlas, questionable point atlas), etc. The content of this part can be directly used for the dual-evaluation of territorial and spatial planning in coordination with the dual-evaluation software system provided by WeChat public number "SPJ_DAS2019".

As the new GC pattern is a brand-new research content in the field of GIS, the research in this field is still in the exploratory stage, and the time is short, bias, and even omissions are inevitable, so please criticize and correct it.

Beijing, China   Wensheng Zhou

Content Summary

In this book, the problems in solving complex GeoComputation(GC) with existing patterns are analyzed, and based on the analysis of the reasons from the point of view of computer programming language, a brand-new GC pattern, Document As a System (DAS), is put forward, and the pattern is expounded from three aspects: theoretical framework, GC language (G language) and the implementation of G language. *The intelligent data processing and analysis system for dual-evaluation of territorial spatial planning* developed with G language is introduced, and the computation model of the main evaluation indexes of dual-evaluation developed with G language is also shown.

This book aims to provide references for the dual-evaluation work carried out at the provincial (regional) and city and county levels in China. At the same time, DAS, a brand-new GC pattern, is also publicized and promoted. I hope DAS can also be applied in other industries involving spatial analysis and spatial data processing, and more researchers are expected to get involved in this field.

This book can be used as a reference manual for management and technical personnel engaged in the dual-evaluation of territorial spatial planning and can also be used as a reference book for students, teachers, researchers, management personnel, and technical personnel of GIS-related majors to learn and study GC language.

Contents

About the Author

Prof. Wensheng Zhou, Ph.D. Deputy Director of the Key Scientific Research Base of the State Administration of Cultural Heritage (Tsinghua University). In 1990, he obtained a master's degree in engineering surveying from Wuhan University of Surveying and Mapping Science and Technology and a doctor's degree in cartography and geographic information engineering from Wuhan University in 2003. In the same year, he entered the School of Architecture of Tsinghua University as a postdoctoral. In 2005, he left the station and stayed at the University. He is mainly engaged in the application of spatial information technology in urban and rural planning and cultural heritage protection. He has presided over or participated in many national, provincial, and ministerial-level scientific research projects, published more than 50 academic papers, and obtained more than 20 software copyrights. He has presided over the compilation of one national standard and four industry standards. The representative achievements include the theory and practice of open WebGIS (Science Press, 2007) and Tsinghua GeoDesign System (ID: 2013sr160411). He participated in the compilation of the technical guidelines for the evaluation of the carrying capacity of resources and environment and the suitability of territorial and spatial development issued by the Ministry of Natural Resources, and as the technical director, he completed the Guangzhou City trial evaluation together with Beijing Tsinghua Tongheng planning and Design Institute Co., Ltd. During this period, a new generation of GeoComputation(GC) pattern, Document As a System (DAS), and GC language (G language) were creatively put forward, and based on this, the "Dual-Evaluation Intelligent Data Processing and Analysis System for Land Spatial Planning" (DAS2019) was developed. The system has been successfully applied to the national "Dual-Evaluation" work jointly presided over by the China Land Survey and Planning Institute, the China Geological Survey and the National Marine Information Center, the trial evaluation of Guangzhou, and the Dual-evaluation work of other provinces, cities, and counties in the country. e-mail: zwsbj@163.com

Part I
The Theory of the New GC Pattern

Chapter 1
Research Background

1.1 Dual-Evaluation of Territorial and Spatial Planning

1.1.1 Territorial and Spatial Planning

Land space refers to the regional space under the jurisdiction of national sovereignty or sovereign rights, and is the place and environment where citizens live. Including territory, airspace, territorial sea, and exclusive economic zone waters stipulated in the *National Convention on the Law of the Sea* (Wu and Pan 2011). Land space is a complex geographical and social space, Including land resources, water resources, mineral resources, marine resources, ecological resources, social and economic resources and other different objects, involving land and natural environment, social and economic environment and psychological and cultural environment, specifically including physical geographical space, social and economic space, psychological and cultural space and network data space (Wu et al. 2019a). From the perspective of human settlement environment science, land space is also human settlement environment, both of which are like two sides of the same silver coin (Wu 2018).

Territorial and spatial planning is the guide for national space development, the spatial blueprint for sustainable development, the basic basis for all kinds of development, protection and construction activities, and the overall arrangement for a national or regional government department to carry out layout and long-term planning of territorial and spatial resources under its jurisdiction. Territorial and spatial planning is a ruler to measure the modernization of the national governance system and governance capability, a carrier platform to ensure the national space competitiveness and sustainable development, and an important tool to safeguard the national space security pattern (Xu et al. 2017).

Minister of Natural Resources Lu Hao pointed out that spatial planning is an integrated work, Land use planning, urban and rural planning, and main functional area planning all provide the basis and experience for spatial planning. However, spatial planning is not the addition or combination of the three, but the new creation based on the three and higher than the three. In this sense, planning is neither urban

© Surveying and Mapping Press 2021
W. Zhou, *A New GeoComputation Pattern and Its Application in Dual-Evaluation*,
https://doi.org/10.1007/978-981-33-6432-5_1

and rural planning nor land use planning, but should be territorial and spatial planning (Lu 2018).

At present, territorial and spatial planning is an important measure to ensure and promote the overall, coordinated and sustainable development of economy and society in our country, and has important historical significance. In September 2015, the CPC Central Committee and the State Council issued the *Integrated Reform Plan for Promoting Ecological Progress*, proposing the establishment of a spatial planning system, requiring the integration of various spatial plans currently prepared separately by various departments, the preparation of unified spatial plans, and the realization of full coverage of the plans. In March 2018, the Central Committee of the Communist Party of China promulgated the *Plan for Deepening the Reform of Party and State Institutions* and set up the Ministry of Natural Resources. It also called for "strengthening the guiding and binding role of territorial and spatial planning on various special plans, promoting the integration of multiple regulations, and realizing the organic integration of land use planning, urban and rural planning, etc." In December 2018, the CPC Central Committee and the State Council issued *Opinions on Unifying the Planning System to Better Play the Guiding Role of the National Development Strategy*, it has further established the position and role of territorial and spatial planning in the new era. Planning should lead economic and social development. On the basis of the planning of the main functional areas, a national planning system with accurate positioning, clear boundaries, complementary functions and unified convergence should be established, which is guided by the national development planning, based on spatial planning, supported by special planning and regional planning, and composed of national, provincial, city and county levels of planning.

The research on territorial and spatial planning has yielded fruitful results, mainly including the discussion based on "three regulations in one" and "multiple regulations in one" (Gu 2015; Zhu et al. 2015; Xie and Wang 2015; Xu et al. 2017), discussion on spatial planning system (Wu 2007; Wu et al. 2019; Lin et al. 2018), discussion on theoretical system of spatial planning (Fan et al. 2014; Hao 2018), spatial planning and spatial governance (Sun 2018; Zhang et al. 2018a, b), reference and enlightenment from international experience in spatial planning (Lin et al. 2011; Cai et al. 2018).

1.1.2   Dual-Evaluation of Territorial and Spatial Planning

The so-called Dual-evaluation of territorial and spatial planning refers to the evaluation of the carrying capacity of resources and environment and the evaluation of the suitability of land space development (hereinafter referred to as Dual-evaluation).

The carrying capacity of resources and environment refers to the natural upper limit, environmental capacity limit and ecological service function bottom line that land space can carry human life and production activities. Carrying capacity evaluation of natural environment is a comprehensive evaluation of natural resources and ecological environment background conditions, reflecting the carrying capacity grade of land space under the direction of urban development, agricultural production and ecological protection functions, and is a prerequisite for carrying out suitability evaluation of land space development.

The suitability of land space development refers to the suitability of land space to different development and utilization modes such as urban development, agricultural production and ecological protection. The suitability evaluation of land space development is based on the carrying capacity of resources and environment, The comprehensive evaluation of the suitability of territorial and spatial development and protection is an important basis for the reasonable delineation of urban, agricultural, ecological space and ecological protection red lines, permanent basic farmland and urban development boundaries, and is also a scientific reference for determining the threshold range of land development intensity and formulating comprehensive control measures (Fan 2018).

The Dual-evaluation of territorial and spatial planning emphasizes and reveals the coordination relationship between resources and environment and human society, which meets the needs of territorial and spatial planning to emphasize the allocation and reconstruction of various elements such as nature, resources and economy, and can directly provide an important basis for the coordination of spatial distribution and allocation of various elements of territorial and spatial planning.

It can be seen from this that the Dual-evaluation can objectively reflect the advantages and disadvantages of the current resources and environment and discover the potential for future development. At the same time, the existing planning results can be corrected by comparing the current situation (such as the current situation of development and utilization, the red line of basic farmland, the red line of ecological protection, etc.). In addition, the results of the Dual-evaluation clearly define the key points of future ecological restoration within the planning scope from the space, which is helpful to analyze the method to continuously improve the carrying capacity of land space, resources and environment.

As the basis of territorial and spatial planning, the Dual-evaluation of territorial and spatial planning is an important measure to implement the national ecological civilization construction strategy, and is also a concrete manifestation of the implementation of relevant national policy opinions and requirements. The *Opinions of the CPC Central Committee and the State Council on Establishing a Territorial and spatial planning System and Supervising Its Implementation* issued on May 10, 2019 clearly states that, "Adhering to the policy of giving priority to conservation, On the basis of the evaluation of the carrying capacity of resources and environment and the suitability of territorial and spatial development, scientific and orderly overall layout of ecological, agricultural, urban and other functional spaces, delineation of ecological protection red lines, permanent basic farmland, urban development boundaries and other spatial control boundaries as well as various sea area protection lines, strengthening the bottom line constraints, and reserving space for sustainable development".

At present, the work of Dual-evaluation is being carried out nationwide, and there are still many disputes over the theory and methods of Dual-evaluation (Fan 2019; Gu 2019; Hao et al. 2019). But with a technical standpoint, the detailed work content of Dual-evaluation of territorial and spatial planning is GIS spatial analysis. By using the spatial analysis tools of GIS, such as coordinate conversion, format conversion, spatial interpolation, network analysis, spatial overlay, spatial statistics, etc., comprehensively analyze land resources, water resources, mineral resources, marine resources, cultural resources, ecology, environment, disasters, location and other factors, and evaluate the carrying capacity of resources and environment and the suitability of territorial and spatial development at different levels of the country, province, city and county.

It can be said that the Dual-evaluation work is a large-scale practice of spatial analysis technology in the whole country, and it is a test of relevant theoretical and technical research for many years.

1.2 Technical Guidelines for Dual-Evaluation

1.2.1 Preparation of Dual-Evaluation Technical Guidelines

To effectively implement the Dual-evaluation work, in November 2018, the Spatial Planning Bureau of the Ministry of Natural Resources issued the notice of *Research on Technical Methods for Evaluation of Carrying Capacity of Resources and Environment and Suitability of Territorial and Spatial Development*. It is proposed to carry out the Dual-evaluation work to provide the basis and support for the preparation of territorial and spatial planning at all levels, and to serve the major decision-making requirements such as population industry and urbanization pattern, space development policy and land consolidation. Carrying out the Dual-evaluation work, scientifically guiding the scientific development of territorial and spatial, rational layout, efficient utilization and orderly management can effectively ensure the scientificity and rationality of territorial and spatial planning, and are also important prerequisites and foundations for realizing coordinated and sustainable regional development.

The main objective of this topic is to compile the *Technical Guidelines for Evaluation of Carrying Capacity of Resources and Environment and Suitability for Territorial and Spatial Development* (hereinafter referred to as *Technical Guidelines*), which will give guidance to the Dual-evaluation of territorial and spatial planning nationwide.

The whole research work is divided into four stages:

1. **Parallel research stage (November 2018–December 2018)**

According to the requirements of the notification, the participating units carry out the research on the carrying capacity of resources and environment and the evaluation technology and method of the suitability of territorial and spatial development, and put forward their respective technical schemes. The research content mainly includes the related theory of Dual-evaluation, the connotation of carrying capacity of resources and environment, the suitability of territorial and spatial development, the technical route and index system of Dual-evaluation and so on.

2. **Formation stage of** *Technical Guidelines (draft for comments)* **(December 2018–January 2019)**

The Spatial Planning Bureau of the Ministry of Natural Resources has set up a Dual-evaluation technical research group to integrate the technical schemes of participating units and form a *Technical Guidelines (draft for comments)* after several rounds of discussions.

3. **Trial evaluation stage (January 2019–April 2019)**

In order to improve *the Technical Guidelines (draft for comments)* and make it better guide the development of Dual-evaluation, the Spatial Planning Bureau of the Ministry of Natural Resources has selected nine regions to carry out the pilot work of Dual-evaluation at provincial or municipal and county levels. The nine regions include Qingdao in Shandong Province, Jiangsu Province, Suzhou in Jiangsu Province, Guangdong Province, Guangzhou in Guangdong Province, Chongqing Municipality, Fuling District in Chongqing Municipality, Ningxia Hui Autonomous Region and Guyuan City in Ningxia.

After nearly three months of trial evaluation work (including on-site research, on-site exchanges, multiple rounds of adjustment of the *Technical Guidelines* and revision of the results of the trial evaluation), the pilot projects successfully completed the trial evaluation work, and held a "Report and Exchange Meeting on the Results of the Trial Evaluation" in the Ministry of Natural Resources on April 2, 2019.

4. **Guidelines improvement stage (April 2019–Early 2020)**

According to the results of the nine regional trial evaluations, the Spatial Planning Bureau of the Ministry of Natural Resources listened to the opinions of experts, further improved *the Technical Guidelines*, and officially issued *the Guidelines for Evaluation of the Carrying Capacity of Resources and Environment and the Suitability of Territorial and Spatial Development (for Trial Implementation)* in January 2020.

1.2.2 Contents of Dual-Evaluation

According to the requirements of *Technical Guidelines (June Edition)*, the technical process of Dual-evaluation of territorial and spatial planning is shown in Fig. 1.1.

The Dual-evaluation work mainly includes the following aspects.

1. **Construction of basic database**

Basic data is the basis of the Dual-evaluation work. Relevant data need to be obtained from relevant departments (such as natural resources department, agriculture department, ecology and environment department, emergency management department,

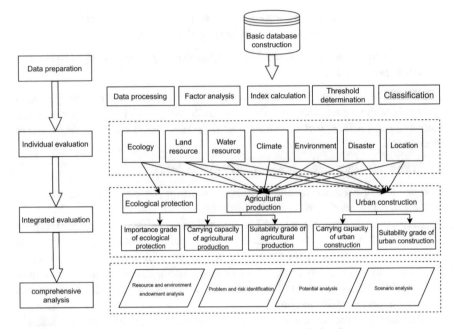

Fig. 1.1 Dual-evaluation technical process of territorial and spatial planning

water sector, meteorological department) according to the evaluation content, and these data need to be standardized and put into storage to form the Dual-evaluation basic database.

2. **Individual evaluation**

Individual evaluation is the basic evaluation content of Dual-evaluation, mainly including 7 individual evaluations of ecology, land resources, water resources, climate, environment, disasters and location. In addition, according to the regional characteristics of each evaluation area, optional evaluation contents such as ocean, cultural protection and mineral resources can be carried out.

3. **Integrated evaluation**

Integrated evaluation is an evaluation based on individual evaluation, oriented to ecological protection, agricultural production and urban construction, and comprehensively considering social and economic factors. It mainly includes ecological protection importance grade, agricultural production carrying scale, agricultural production suitability grade, urban construction carrying scale and urban construction suitability grade.

4. **Comprehensive analysis**

Comprehensive analysis is based on individual evaluation and integrated evaluation, Based on the current situation of land use, ecological red line, permanent agricultural farmland protection red line, urban development boundary and other contents, the characteristics of resources and environment endowment in the evaluation area are analyzed, problems and risks are identified, agricultural production and urban construction potential are analyzed, and scenario analysis is carried out according to the future development of the evaluation area.

5. **Compilation of report about dual-evaluation**

Report According to the GC results of the Dual-evaluation and the style stipulated in *Technical Guidelines*, the Dual-evaluation report is compiled from the aspects of resource and environment endowment characteristics, problem risk identification, ecological protection importance evaluation, agricultural production function orientation evaluation, urban construction function orientation evaluation, scenario analysis and relevant suggestions.

1.3 Implementation of Dual-Evaluation

1.3.1 Requirements for Dual-Evaluation

As the basis of territorial and spatial planning, the quality of Dual-evaluation has an important impact on the compilation of territorial and spatial planning and is a serious and scientific work. Therefore, the development of Dual-evaluation should follow the following principles.

1. **Scientific, systematic and normative**

The Dual-evaluation work is a kind of landing work carried out at different levels in the country, not general academic research. For this reason, in addition to formulating *Technical Guidelines* to guide the work, the Dual-evaluation work also needs to consider how to implement the work. Formulating a scientific, systematic and standardized specific practice route is an important guarantee for the smooth implementation of the Dual-evaluation work.

2. **The quality of evaluation results can be controlled, checked and traced back**

From a technical point of view, the specific content of Dual-evaluation involves a large amount of spatial analysis content. Due to the large number of GC steps, it is easy to cause errors in evaluation results. Therefore, the evaluation process needs a controllable, verifiable and traceable technical means to effectively ensure the quality of the Dual-evaluation results.

3. **Easy implementation of evaluation**

The Dual-evaluation work needs to be carried out at all levels of the country, provinces, cities and counties. Considering the differences in data and technical personnel reserves, the Dual-evaluation work needs to consider the issue of enforce ability. Simple and easy technical methods should be provided as much as possible to ensure the smooth implementation of the Dual-evaluation work.

1.3.2   Challenges of Dual-Evaluation

Judging from the Dual-evaluation work described above, the Dual-evaluation work is a large-scale GC project mainly based on quantitative spatial analysis in fact, involving a large amount of spatial analysis content, and the work is complicated and arduous, which is embodied in the following three aspects.

1. **Extensive implementation scope**

The Dual-evaluation involves the work at all levels of the country, provinces, cities and counties. The basic conditions (mainly refers to data and technical conditions) for carrying out this work vary greatly from place to place. For regions with high level of information development and economic development, such as Guangdong Province and Jiangsu Province, their data base and talent reserve can generally meet the requirements of Dual-evaluation. On the contrary, it is difficult and challenging to complete the Dual-evaluation work with good quality and quantity in regions with poor data base and talent reserve.

2. **Involving many specialties**

According to the requirements of *Technical Guidelines*, the individual evaluation of Dual-evaluation includes the evaluation contents of climate, land resources, water resources, environment, ecology, disasters and location, while the optional evaluation includes the contents of ocean, cultural protection and mineral resources. These evaluations involve ecology, environment, geological disasters, geographical science, hydrology and water resources engineering, environmental science, ecology, geotechnical engineering, engineering geology, ocean, urban and rural planning and other professional contents, which require a certain understanding and knowledge of relevant specialties in order to be competent for the Dual-evaluation work.

3. **Difficulty in data processing**

The Dual-evaluation work involves a large amount of spatial data processing. The *Technical Guidelines (June edition)* includes 7 individual evaluations, 5 integrated evaluations and 5 comprehensive analyses, in addition to 3 optional evaluations. For these evaluations or analyses, they all need to go through multiple spatial data processing steps, ranging from a few to dozens. According to our statistics, the whole Dual-evaluation work requires more than 400 processing steps. Figure 1.2 shows the number of operations for each GC task. If conventional technical means are used to complete these spatial analysis works, it is a definite challenge for technicians.

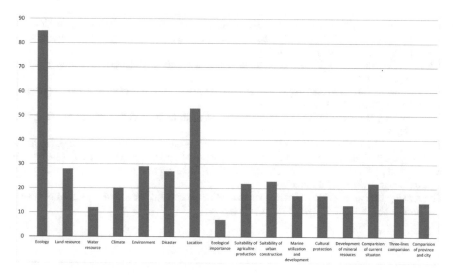

Fig. 1.2   Dual-evaluation GC tasks analysis

References

Cai Y, Gordon JA, Xie X (2018) Experience and enlightenment of spatial planning systems in major developed countries. Land China 5:28–30 (in Chinese)

Fan J (2018) Guidelines for evaluation methods of carrying capacity of resources and environment and suitability of land and space development. Science Press, Beijing (in Chinese)

Fan J (2019) "Dual-evaluation" of territorial and spatial planning-theory and method [EB/OL]. https://www.planning.org.cn/report/view?Id=334.2019-10-28 (in Chinese)

Fan J, Jiang Z, Chen D (2014) Scientific basis and practical strategy of coordinated spatial layout planning. Urban Plan 38(1):16–25 (in Chinese)

Gu C (2015) On the separation, evolution and fusion of China's "multiple rules." Geograph Res 34(4):601–613 (in Chinese)

Gu C (2019) Scientific "dual-evaluation" is the key and foundation of land and space planning in the new era [EB/OL]. https://www.sohu.com/a/322995265_656095 (in Chinese)

Hao Q (2018) Reflections on the construction of spatial planning system in the context of institutional reform. Geograph Res 37(10):1938–1946 (in Chinese)

Hao Q, Deng L, Feng Z (2019) Reflection on carrying capacity in territorial and spatial planning: concept, theory and practice. J Nat Resour 34(10):2073–2086 (in Chinese)

Lin J, Chen X, Wei X (2011) Discussion on coordination of spatial planning in China-international experience and enlightenment of spatial planning. Mod Urban Res 26(12):15–21 (in Chinese)

Lin J, Wu Y, Wu J et al (2018) On the construction of spatial planning system-also analyzing the relationship between spatial planning, land and space use control and natural resources supervision. Urban Plan 42(5):9–10 (in Chinese)

Lu H (2018) Planning is neither urban and rural planning nor land use planning [EB/OL]. https://www.sohu.com/a/233233000_275005 (in Chinese)

Sun A (2018) Reflections on spatial planning reform. J Urban Plan 1:10–17 (in Chinese)

Wu T (2007) Prospect of China's regional spatial planning system in the new era. Urban Plan 07:39–46 (in Chinese)

Wu L (2018) Spatial planning system reform and human settlements science development [EB/OL]. https://www.sohu.com/a/278065942_611316 (in Chinese)

Wu C, Pan W (2011) Theory and method of land and resources. Science Press, Beijing (in Chinese)

Wu T, Lu Q, Zhou W et al (2019) On the construction of territorial and spatial planning system. Res Urban Reg Plan 11(1):1–12 (in Chinese)

Wu C, Ye Y, Wu Y et al (2019) Territorial and spatial planning. Geological Publishing House, Beijing (in Chinese)

Xie Y, Wang W (2015) From "integration of multiple rules" to reconstruction of spatial planning system. J Urban Plan 03:15–21 (in Chinese)

Xu J, Shen C, Hu T et al (2017) General ideas and main tasks of constructing China's spatial planning system. Planner 33(02):5–11 (in Chinese)

Zhang B, Lin Y, Liu W et al (2018a) Urban development boundary and national space governance-ideological basis for delineating urban development boundary. J Urban Plan 04:16–23 (in Chinese)

Zhang J, Lin H, Hao C (2018b) Changes and reform of China's spatial planning system in 40 years. Econ Geogr 38(7):1–6 (in Chinese)

Zhu J, Deng M, Pan A (2015) "Three rules in one": exploring the order and regulation of spatial planning. Urban Plan 39(1):41–47 (in Chinese)

Chapter 2
Relevant Research

2.1 GeoComputation

2.1.1 Spatial Analysis

With the growing maturity of GIS technology and the arrival of the era of big data, the field of GIS has gradually shifted from focusing on the construction of various spatial databases and the development of GIS application systems to focusing on spatial analysis and spatial modeling to solve geo-spatial problems and provide services for people's work and life.

The concept of spatial analysis (SA) originated from the metrological revolution of geography and regional science in the 1960s. As the essence of geography, spatial analysis is data analysis and mining to solve geographical spatial problems. Goodchild once pointed out that "the real function of geographic information system lies in its use of spatial analysis technology to analyze spatial data". Unlike general spatial databases, information systems and cartographic systems, GIS spatial analysis can not only manage massive spatial data, query, retrieve and measure information, but also the hidden modes, relationships and trends in geo-spatial numbers can be analyzed through graphic operation and data simulation operation, and information with guiding significance for scientific decision-making can be mined, thus solving complex geo-science application problems and carrying out geo-science comprehensive research.

Because the content of spatial analysis is very complex, the understanding of spatial analysis in GIS academic circles are also different. The following are some viewpoints of famous scholars: spatial analysis is a statistical description or explanation of spatial information, attribute information or the information in common with both (Goodchild 1987); Spatial analysis is a quantitative study of geo-spatial information. Its conventional ability is to manipulate spatial data into different forms and extract its potential information (Openshaw and Openshaw 1997; Bailey and Gatrell 1995); Spatial analysis refers to obtaining derived information and new knowledge from the spatial relationship between GIS targets (Li

© Surveying and Mapping Press 2021
W. Zhou, *A New GeoComputation Pattern and Its Application in Dual-Evaluation*,
https://doi.org/10.1007/978-981-33-6432-5_2

and Cheng 1995); Spatial analysis refers to spatial data analysis technology based on the location and shape of geographical objects. Its purpose is to extract and transmit spatial information (Guo 1997).

In general, spatial analysis is a technology which integrates spatial data analysis and spatial simulation. It digs potential spatial information through GeoComputation and spatial expression to solve practical geographical problems, its fundamental goal is to establish an effective spatial data model to express the spatio-temporal characteristics of geographical entities, develop application-oriented spatio-temporal analysis and simulation methods, and dynamically and globally describe the spatial distribution relationship between geographical entities and geographical phenomena in a digital way, thus reflecting the inherent laws and changing trends of geographical entities (Liu et al. 2005).

Spatial analysis plays an important role in the research of geography, government decision-making, commercial and economic decision-making, and public travel. For the study of geography, it can analyze the distribution law of various geographical phenomena, reveal the correlation and temporal and spatial evolution of geographical matters, analyze the spatial structure of geographical matters and reveal the spatial effect of geographical things. For government decision-making, it can provide analysis tools and decision support for government management and scientific basis for specialized management of departments. For commercial economic decision-making, commercial geography analysis, commercial location analysis and commercial marketing auxiliary decision-making can be carried out. For public travel, it can provide support in vehicle navigation, travel decision-making and safe driving (Cui 2019).

2.1.2   GeoComputation

Although spatial analysis provides an effective means to solve geo-spatial problems, due to the large amount of geo-spatial data and the complexity of geographic problems, the concrete implementation of spatial analysis is in trouble. For this reason, Openshaw and Abrahart (1996) of the School of Geography of Leeds University put forward the concept of GeoComputation (GC) in 1996 and published the first monograph of GC, *GeoComputation: A Primer*, in 1998. Openshaw and Abrahart (1996) believed that the application of geography to computers is different from GIS as an auxiliary computation tool during the metrological movement period, but a scientific research method. The proposal of GC no longer only regards GIS as a system, but focuses not on how to better collect, acquire, manage and transmit geographic spatial data, but on using advanced computation technologies, including various advanced algorithms and high-performance computation, to solve geographic problems.

In 1996, the first annual meeting of GC was held at Leeds University, marking the beginning of GC as a discipline. The main topics of this conference include spatio-temporal dynamics, application of high-performance computer technology in geography, interoperability and GC, intelligent self-agent, neural network and fuzzy computation, spatial theory and spatial logic, inference and mode mining, real environment and virtual environment, interactive visualization and application of GC. Since then, more than 10 international conferences have been held in the world. Since this century, GC and GIScience International Conference have been held alternately every other year, which has jointly promoted the development of GC. In recent years, the GC Conference has focused on geographic data analysis, geographic system simulation, CA, ABS, spatial operations, geographic research visualization, automatic mapping, geographic information organization, etc. (Cai 2011).

At present, there is no unified definition of the connotation of GC. Longley et al. defined GC as the application of computationally intensive methods in natural and human geography (Longley et al. 2001). Rees et al. proposed that "GC can be defined as a method of dealing with geographical problems when computation technology is applied" (Rees and Urton 1998). Openshaw and Abrahart (2000) thinks that GC refers to the application of computational science paradigm in a wide range of problem research under the background of geography and earth system. It is a method based on high-performance computation to solve usually unsolvable or even unknowable problems (Openshaw and Abrahart). Gahegan (1999) pointed out that "GC is a conscious effort by human beings to explore the relationship between geography and computer science. It is a real quantitative geography technology and a rich source for computer scientists to carry out computational applications."

After years of research, Chinese scholars Wang et al. (2007) and Wang and Wu (2011) believe that, in a broad sense, GC is a branch of geography that uses computer methods as basic scientific tools to process geographic information and analyze geographic phenomena. It includes geographic information processing and management, geographic data mining, geographic process modeling and simulation, as well as software engineering and computation system research supporting these processing and analysis, such as geographic information system, geographic decision support system and spatial grid system. In a narrow sense, GC is one of the core contents of geographic information science. It mainly studies methodological issues of geographic information science, including algorithms, modeling and computation systems.

To sum up, GC is to use various types of geographic and environmental data to develop relevant computation tools in the whole system of computer science methods, focusing on the methodological issues of spatial analysis, including algorithms, modeling and computation systems.

In recent years, with the advent of the era of big data, the rapidly growing geographic big data has become an important part of big data flow. The further development of spatial big data will also effectively resolve the long-standing data bottleneck problem in the field of geographic information. In addition, with the development of hardware technologies such as high-performance computers and software technologies such as cloud computation and mobile computation, people will pay more attention to the research of GC (Cui 2019).

2.1.3 GC Pattern

At present, in the research of GC, people pay attention to the research of theories, methods and systems, but there is less discussion on how to apply these methods to solve specific geographical problems. But for the business and research related to geographical analysis, how to construct and compute geographical models in an orderly, standardized and effective way is the more concerned content. Here, the author attributes it to the problem of GC pattern. The so-called GC pattern is the method or means used for GC. Summing up the existing GC methods, there are no more than two types: manual pattern and automation pattern. Among them, automation pattern is divided into visual modeling pattern, script modeling pattern and independent system development pattern according to the degree of professional requirements for computer programming.

1. **Manual pattern**

Manual pattern, or toolbox pattern, is currently a commonly used method for geographic analysis. It completes specific GC tasks through a series of GIS analysis tools provided by GIS platform (such as overlay analysis, buffer analysis, network analysis, spatial interpolation, kernel density analysis, etc.). For example, more than 800 processing tools are provided in the ArcToolbox of ESRI ArcGIS version 10.1, mainly including format conversion tools, data management tools, mapping tools, spatial analysis tools, three-dimensional analysis tools, network analysis tools, spatial tools, etc. (as shown in Fig. 2.1).

2. **Visual modeling pattern**

Visual modeling pattern uses visual methods to help users build geographic analysis models, thus realizing the automation of GC. Because this modeling method does not

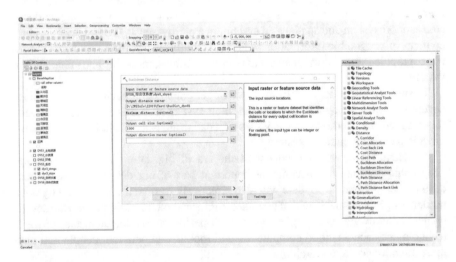

Fig. 2.1 Toolbox pattern in ArcGIS

require users to know programming knowledge, it reduces the difficulty of geographic analysis model development. For example, Model Builder in ESRI ArcGIS is a visual modeling tool, which is based on the basic principle of graphic modeling and uses intuitive graphic language to express the problems to be studied with one or more specific process models. In the modeling process, graphics can be defined to represent input data, output data and spatial processing tools respectively. These graphic elements are combined in the form of flow charts and perform spatial analysis operations. In Model Builder, you can drag various geographic processing tools and data sets to be processed in the toolbox into the Model Builder interface, and then connect them in an orderly way. At the same time, models and scripts can be combined and applied. Simple processes can be directly modeled by Model Builder through visualization. Complex processes can be programmed by script language. Models can directly call these results. Figure 2.2 is an example of a geo-processing model built with Model Builder.

3. **Script development pattern**

Script development pattern uses proprietary scripting language (such as MapBasic language in MapInfo) or general scripting language (such as VBA and Python) to call the encapsulated algorithms in GIS platform to complete complex data processing and GC model construction. ESRI has supported a variety of scripting languages since ArcGIS 9, including Python, VBScript, JavaScript, Jscript, and Perl. Although system languages (such as C++ and .NET) can also build GC models and realize the automation of GC tasks, scripting languages are more concise and therefore are favored by many users. Figure 2.3 is an example of the plot ratio computation model written by the author using ArcPy in ArcGIS.

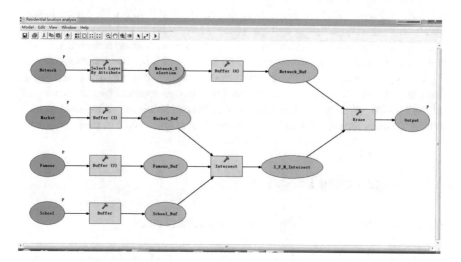

Fig. 2.2 Example of visual modeling

Fig. 2.3 Python script development example

4. **Independent system development pattern**

The independent system development pattern uses system languages (such as C++ and .NET) to develop independent GIS application systems from the bottom or on the basis of GIS components. At present, component GIS development pattern is widely used (Song 1998; Zhou et al. 2007). The commonly used GIS components include MapX of MapInfo Company and ArcEngine of ESRI Company in the United States, and SuperMap iObjects of SuperMap Company in China. ArcEngine is a newly added product in ArcGIS 9.0 series, which consists of ArcEngine Developer Kit and ArcEngine Runtime. It is an embedded GIS software with complete class library. It supports multiple languages (COM, JAVA, .NET, and C++) and multiple systems (Windows and Unix). Developers can not only customize complete GIS software through ArcEngine, but also embed GIS functions into other existing software. Figure 2.4 is an auxiliary planning system developed by the author using ArcEngine. It should be noted that even if the component GIS development pattern is adopted, it is still a difficult thing for a technician who has not received professional training to develop an independent system due to the basic requirements of programming.

2.2   Programming Language

2.2.1   *Development of Programming Languages*

Automated GC pattern is the development direction of GC pattern, and programming language is the basis to achieve this goal. Therefore, it is necessary to briefly sort out the relevant research of programming language here.

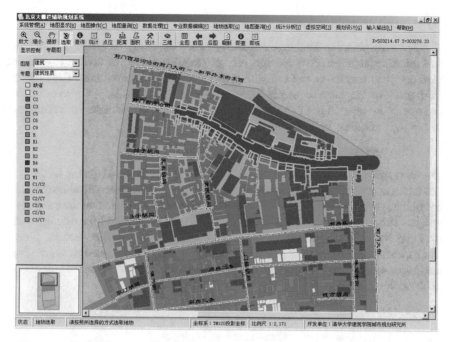

Fig. 2.4 Example of independent system development

1. **Development of programming languages**

Language is a tool for human beings to exchange ideas and information, express feelings in the long-term historical development process. This kind of language, such as Chinese, English and French, is commonly called natural language. In addition, people have created some languages for some purpose, such as semaphore and dumb language. This kind of language is usually called artificial language. As a communication tool between human beings and computers, programming language belongs to artificial language, because it is designed and created by human beings, rather than a natural language formed by human beings in the long-term communication process, and can be simply understood as a language that can be recognized by both computers and human beings.

Since the first computer appeared in 1946, hundreds of programming languages have been produced to adapt to different uses. These programming languages have gone through three stages of development (Wang et al. 2015).

(1) **First generation language**

The first generation of languages, commonly known as machine languages, is represented by binary zeros and ones. It is an instruction system that depends entirely on machines. Programmers program by punching holes in paper tape. This language is difficult to write and understand, and few people can master it.

(2) Second generation language

The second-generation language is usually called assembly language. It symbolizes machine language. Programmers can use some symbolic instructions to operate memory, thus improving the writability and readability of programs. The change from machine language to assembly language is a qualitative leap, which not only represents the separation of software and hardware, but also enables programmers to express their ideas in their own language.

(3) Third generation language

The third-generation language usually refers to the advanced programming language. It is a programming language close to natural language and mathematical formula. It has nothing to do with the hardware structure and instruction system of the computer and has stronger expression ability. It can conveniently represent the operation of data and the control structure of the program, can better describe various algorithms, and is easy to learn and master. There are hundreds of advanced programming languages so far. Table 2.1 lists the basic information of the main advanced languages in chronological order.

Table 2.1 Basic information table of main advanced languages

Language	Age	Creator	Application field
FORTRAN	1954–1957	J. Backus (IBM)	Numerical computation
ALGOL60	1958–1960	Committee	Numerical computation
COBOL	1959–1960	Committee	Business data processing
LISP	1956–1962	J. McCarthy (MIT)	Symbolic computation
ALGOL68	1963–1968	Committee	For all purpose
BASIC	1964	John G. Kemeny, Thomas E. Kurtz (Dartmouth College)	System program
Pascal	1971	N. Wirth (Zurich ETH)	General and structural programming
Prolog	1972	A. Colmerauer (Marseille, France)	AI
C	1974	D. Ritchie (Bell Laboratory)	System program
Smalltalk	1971–1980	A. Kay (Xerox PARC)	Personal computation environment
C++	1984–1989	Bjarne Stroustrup	System program
Python	1989	Guido van Rossum	Application system
Java	1991–1994	James Cosling (Green group)	Web browser
Delphi	1989	Anders Hejlsberg	Graphical interface, visualization

(continued)

Table 2.1 (continued)

Language	Age	Creator	Application field
R	1991	Ross Ihaka, Robert Gentleman (University of Auckland, New Zealand)	Statistical analysis
JavaSript	1995	Netscape Company	Dynamic web page making
PHP	1995	Rasmus Lerdorf	High dynamic web page making
D	1999	Digital Mars Company	System program
C#	2000	Anders Hejlabeerg	Application system
Go	2009	Rob Pike	Application system

2. Classification of advanced programming languages

At present, there are dozens of popular advanced programming languages, but they can be broadly divided into four categories:

(1) Imperative language. This programming language is an action-based language with von Neumann's computer architecture as its background. Machine language and assembly language are the earliest imperative languages. In this language, the computer is regarded as a sequence of actions, and the program is an operation sequence written with the operation commands provided by the language. Writing a program with imperative programming language is to describe the process of each step in the process of solving a problem. The running process of the program is the process of solving the problem, so it is also called procedural language. Most popular languages are of this type, such as Fortran, Pascal, Cobol, C, C++, Basic, Ada, Java, C #, etc. In addition, various scripting languages are also considered of this type.

(2) Functional language. This is a non-Von Neumann programming language. Its main components are original functions, definition functions and function types. This language has a strong ability to organize data structures and can treat a certain data structure (such as an array) as a single value. The function can be taken as a parameter, and its result can also be a function. This defined function is called a high-order function. The program is a function. The program acts on structural data to produce structural results, which fundamentally changes the "word-by-word" working mode of von Neumann language. This language is very suitable for Computation of artificial intelligence and other work. Typical functional languages such as Lisp, Haskell, ML, Scheme, F #, etc.

(3) Logical language. This is a logical programming language oriented to deductive reasoning. Its semantic basis is a formal logic system based on a set of known rules. This language is mainly used in the implementation of expert systems. The most famous logical language is Prolog.

(4) Object-oriented language. This language is a kind of programming language that takes objects as the basic program structure unit. The design used to describe

takes objects as the core, and objects are the basic components of the program running time. The language provides such components as class and inheritance, which have four main characteristics: recognition, polymorphism, class and inheritance. At present, the mainstream programming provides object-oriented support, but some languages are directly based on the basic object-oriented model. The syntax form of the language is the basic object operation, and the main pure object-oriented language is Smalltalk.

3. **Development direction of programming language**

With the birth of computer programming language, the computer field has been further developed. In its continuous development and innovation, it has partially met the application needs of all sectors of society. However, the phenomenon that software is in short supply is still very serious. The contradiction between the increasing personalized needs of users and the shortage of software products is getting worse and worse. The reason for this phenomenon is that the current programming language still cannot meet people's needs. The development direction of programming language should be gradually close to semantics and closer to human natural thinking. With the development of the first generation of languages, programming languages are more and more suitable for programmers and improve their work efficiency. In other words, the technical threshold for system development has been gradually lowered. This is mainly reflected in the gradual improvement of method encapsulation and the process of writing programming languages closer to natural languages and technical documents. When software develops to a certain extent, the future language is a programming language equivalent to natural language, thus realizing everyone's programmability and highly encapsulated methods. Compilers can recognize people's natural language semantics and convert them into machine-recognizable languages. All people need to do is translate the requirements into standardized requirements analysis documents, which is the highest stage of software development.

2.2.2 Thinking About Programming Languages

With the development of programming language up to now, although great progress has been made from the first generation of machine language to the present advanced language, it should also be noted that only a few people can truly master programming skills to solve problems. It is still a very difficult thing for most people to develop the systems they need in programming languages to solve problems in work, study and life. This shows that the popularization of programming languages has a long way to go.

Analyzing the deep-seated reasons, we will find that the current programming language is still a kind of machine thinking language, and its basic feature is strict logic, while the natural language of people's daily communication is not strict although logical. For this reason, for beginners of programming, they must learn and train a lot to transform natural thinking into machine thinking. That is to say, they need to learn a specific program language, remember some fixed and strict grammar rules, learn to use software engineering methods to convert actual requirements into program descriptions, and need to consider how to design the system, how to select components, and how to make different components work in coordination. This change from natural thinking mode to machine thinking mode is very challenging for most people. This leads to the fact that although most users have been trained for a period, they still cannot develop systems and solve problems through programming languages.

In summary, the current programming language is still a machine thinking language. If this situation is not changed completely, programming is still a distant dream for most people.

2.2.3 End-User Programming

To change the difficulties encountered by the above programming languages, people have made various efforts and attempts and achieved some gratifying results. The end-user programming introduced below has important enlightenment for the research in this paper.

The so-called end-user programming is to let users without software development foundation use their own domain knowledge to develop software. End users are both developers and users. They have the clearest needs for themselves and there will be no communication problems. At the same time, the program developed by end-user programming can be understood and understood by the end-user himself. In the process of software maintenance, the end-user can also understand the shortcomings of the whole software and shorten the maintenance period. In addition, in the process of traditional software development, the shortage of software is another long-term problem. The speed and productivity of software development will never meet the needs of users. However, because end-user programming enables more users to develop their own software, it will improve the speed of software development and shorten the cycle (Burnett et al. 2004).

At present, there are four methods for end-user programming: Program Synthesis, Model Driven Development, Sloppy Programming and Domain-Specific Language (Yu 2013). Among them, Domain Specific Language (DSL) is a kind of programming language with limited expressiveness for a specific domain (Fowler 2010), and its basic idea is "seeking specialization but not perfection". Different from general programming languages, its target range is not to cover all software problems, but a computer language specifically aimed at a specific problem. The primary purpose of domain-specific languages is to bring programs as close to problems in the business domain as possible, thus eliminating unnecessary indirection and complexity. Because domain-specific language is specific to specific domains, it can describe the intention of requirements more concisely, clearly and systematically, thus improving development efficiency and making the development process easier. Fowler (2010) summarizes four important features of domain-specific languages.

First, DSL is a computer programming language. It is a command designed by human beings for computers to execute.

Therefore, in addition to being easy to understand, it must be enforceable. Second, DSL can express language. Grammar cannot be a single expression, but must have a certain ability to express combination.

Third, DSL has limited expression ability. It does not need as strong expression ability as general programming languages (such as C and Python), but only needs it to support a specific field.

Fourth, DSL only focuses on a specific area. Only by paying attention to a specific field can limited expression ability play its greatest role.

The most obvious character of DSL is that the language is a descriptive language, which only explains that what is wanted, instead of writing down in detail how to do and implement it step by step like common programming languages (such as C and Python).

Debasish Ghosh summarizes two differences between domain-specific languages and common programming languages.

First, DSL provides users with a higher level of abstraction, which makes users not care about details such as special data structures or low-level implementations, but only start to solve current problems.

Second, DSL provides a limited vocabulary for its areas of concern. It does not need to provide additional help for specific modeling areas like common programming languages.

These two points make DSL more suitable for non-programmer domain experts. It is precisely these characteristics that DSL can intuitively express application problems at a higher level of abstraction, complete requirement confirmation at the domain level, and realize effective reuse. At the same time, through tool support and domain knowledge reuse, It enables many conversion tasks from specification to executable code to be automated, enables advanced technologies to be used in a wider range, and avoids great ups and downs caused by differences in the skill levels of developers, thus significantly improving the efficiency and quality of software development, accelerating the speed of product development, and meeting the changeable requirements of applications.

Research on DSL has been going on for a long time, and some research results have been obtained, such as HTML describing Web pages, Ant, RAKE, MAKE constructing software systems, BNF paradigm expressing syntax, YACC, Bison, ANTLR generated by parser, SQL for database structured queries, RSpec, Cucumber for behavior-driven testing, CSS describing style sheets, etc. (Debasish 2011).

References

Bailey TC, Gatrell AC (1995) Interactive spatial data analysis. Wiley, New York

Burnett M, Cook C, Rothermel G (2004) End-user software engineering. Commun ACM 47(9):33–37

Cai D (2011) GeoComputation mode in grid environment. Electronic Industry Press, Beijing (in Chinese)

Cui T et al (2019) Principles of geospatial analysis. Science Press, Beijing (in Chinese)

Debasish G (2011) DSLs in action. Manning Publications, Greenwich

Fowler M (2010) Domain-specific languages. Addison-Wesley Professional, New Jersey

Gahegan M (1999) Guest editorial: what is GeoComputation. Trans GIS 3(3):203–206

Goodchild MF (1987) A spatial analytical perspective of geographic information system. Int J Geogr Inf Syst 4:327–334

Guo R (1997) Spatial analysis. Wuhan University of Surveying and Mapping Press, Wuhan (in Chinese)

Li D, Cheng T (1995) Discovering knowledge from GIS database. J Survey Map 24(1):37–44 (in Chinese)

Liu X, Huang F, Wang P (2005) Principles and methods of GIS spatial analysis. Science Press

Longley PA, Goodchild MF, Maguire DJ et al (2001) Geographic information systems and science. Wiley, New York

Openshaw S, Abrahart RJ (1996) GeoComputation. In: 1st international conference on GeoComputation. University of Leeds, London, pp 665–666

Openshaw S, Abrahart RJ (2000) GeoComputation. Taylor & Francis, New York

Openshaw S, Openshaw C (1997) Artificial intelligence in geography. Wiley, London

Rees P, Urton I (1998) GeoComputation: solving geographic problems with new computation power (Guest editorial). Environ Plan A 30:1835–1838

Song G (1998) Research on component geographic information system. Doctoral dissertation of Institute of Geography, Chinese Academy of Sciences, Beijing (in Chinese)

Wang Z, Jing Wu (2011) Computational geography. Science Press, Beijing (in Chinese)

Wang Z, Sui W, Yao Z et al (2007) GeoComputation and its frontier issues. Adv Geogr Sci 26(4):1–10 (in Chinese)

Wang X et al (2015) Programming language and compiler-language design and implementation. Electronic Industry Press, Beijing (in Chinese)

Yu T (2013) Research on domain-specific languages for end users. Shanghai Jiaotong University, Shanghai (in Chinese)

Zhou W, Mao F, Hu P (2007) Theory and practice of open WebGIS. Science Press, Beijing (in Chinese)

Chapter 3
The New GeoComputation Pattern

3.1 Analysis of Existing GeoComputation Patterns

3.1.1 Problem Analysis

The effective implementation of spatial analysis depends on the selection of appropriate GC pattern. However, the existing GC patterns, whether toolbox pattern, visual modeling pattern, script development pattern or independent system development pattern, are difficult to effectively deal with the complex GC task of Dual-evaluation. The specific analysis is as follows.

1. **Toolbox pattern**

For the toolbox pattern, although the toolbox provides a lot of convenience for geographic analysis, it will be found that the toolbox pattern is a very hard-working pattern for technicians when performing more complex GC through the toolbox. In this computation pattern, the technician first needs to find the required tools in the toolbox, and then needs to enter several parameters in the dialog box. Because a GC task is often composed of multiple GC items, and each GC item has a back-and-forth correlation, that is, the input parameters of the latter GC item are the GC results of the former GC item. In this way, when there is a mistake in the GC process (such as improper tool selection or wrong parameter setting), the final GC result will be affected. When a GC task involves fewer GC items, this computation pattern can barely complete the GC task, but when the GC task involves more GC items (for example, the ecological evaluation in the Dual-evaluation involves more than 80 GC items), and it is difficult to ensure the quality and efficiency of the computation work.

In addition, because the parameters in each input toolbox dialog box cannot be retained, many inputs need to be made every time the computation is carried out, and the quality and efficiency of the GC results cannot be guaranteed because the input parameters cannot be guaranteed to be correct.

© Surveying and Mapping Press 2021

W. Zhou, *A New GeoComputation Pattern and Its Application in Dual-Evaluation*,
https://doi.org/10.1007/978-981-33-6432-5_3

2. Visual modeling pattern

As a technique of automated spatial data processing, although there is no need for programming, this pattern is feasible and effective only in relatively simple geographic modeling. When more complex geographic models need to be built, the geographic processing process needs to be converted into hundreds of graphic elements. It is often more difficult to understand such complex graphics in mind than to read the same code. Of course, in this case, "blocks" can be used to represent more complex operations, so that each visual element can represent a large piece of code, but this kind of processing will virtually increase the difficulty of understanding and debugging the model. Figure 3.1 is the plot ratio computation model constructed by the author. This model is only composed of 12 tools. For a model composed of dozens or even hundreds of tools, it is quite difficult to construct, debug and maintain.

3. Script development pattern

In script development pattern, although scripting languages (especially Python) are more acceptable than traditional programming languages, and the code is visible to the user, the scripting language is also a programming language after all. With the characteristics of a general programming language, it is, after all, few people who can truly master this language. For ordinary technicians who have not learned programming, they need a certain amount of learning and training to gradually transform natural thinking into machine thinking to understand, maintain or develop geographical models themselves.

4. Independent system development pattern

Compared with the script development pattern, the system development pattern needs professional developers to develop, and the development cost and maintenance cost

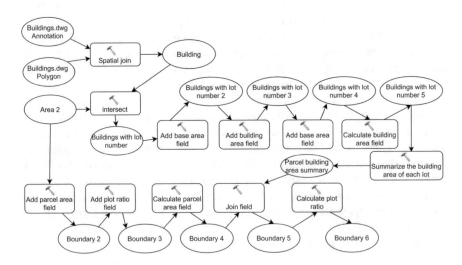

Fig. 3.1 Visual modeling example

are higher. Moreover, because users cannot modify the program code by themselves, the developed system is difficult to meet the flexibility requirements of GC.

In addition to the problems existing in the above-mentioned existing GC patterns, how to effectively test the quality of the calculation results in complex GC, and the existing GC patterns cannot give an effective solution.

3.1.2 Cause Analysis

Since the birth of Geographic Information System (GIS) technology for more than 50 years, although its application in visual management and public service is relatively mature, there are two deficiencies in complex GC:

First, due to nonstandard computation and opaque GC process, the quality of GC results cannot be effectively verified.

Second, in scientific research institutions or enterprises and institutions, general technicians or researchers without programming experience cannot easily build, maintain and use GC models on their own.

The fundamental reason for these problems lies in the lack of a set of systematic, scientific and standardized means and methods to ensure the efficient and high-quality implementation of GC tasks in both engineering and scientific research fields. At present, most GC basically adopts toolbox pattern. As has been analyzed earlier, the biggest problem of toolbox pattern is low efficiency and the quality of results is difficult to be guaranteed. In addition, because the GC results are completely separated from the computation theory and the GC process, it is difficult to carry out the inspection and backtracking of the GC results.

Further analysis of the deep-seated reasons will reveal that most technicians engaged in geographical analysis generally lack programming ability. Programming, even script programming, is an insurmountable hurdle for technicians who have not received a lot of programming training. According to the design logic of current programming languages, only a small number of people can truly master programming skills.

In order to get rid of the dilemma faced by the existing GC pattern, the author tries to construct a new GC pattern to ensure the scientific, systematic, standardized and effective implementation of GC.

3.2 The New GC Pattern

3.2.1 Design Concept

The design of the new GC pattern is based on the following considerations.

1. **Systematic, scientific and normative**

In order to ensure the systematicness and standardization of GC, the new GC pattern should effectively integrate the theory, GC process and GC results of the GC models, to facilitate the inspection and repeated computation of GC results. Traditional computation pattern, whether it is toolbox pattern, visual modeling pattern, script development pattern or independent system development pattern, it is difficult to evaluate the quality of the computational results and to interpret the GC results because of the separation of the theory, GC process and GC results of the GC pattern. This situation exists in both the business departments related to GC and scientific research fields.

2. **Popular graphic editing environment**

Both MS Word and Kingsoft WPS are word processing software used by people in daily life and have a wide range of users. In addition, they not only have strong graphic editing ability, but also have strong automatic processing ability, which provides the possibility to realize the automation of GC processes. It is completely feasible to use them as containers for the theory, GC process and GC results of GC models.

3. **Popular programming language**

Generally, the automatic implementation of GC is realized through programming, but considering the reality that most people do not have programming ability, it is necessary to design a new programming language that is easy for the public to accept. Because the design and implementation of general programming language is a difficult task, it requires a lot of manpower and material resources. However, the domain-specific language introduced in section 2.2.3 provides a new idea for popular programming languages, that is, "seeking specialization but not perfection". Its target range does not cover all software problems, but is a computer language specially aimed at GC tasks to eliminate the complexity of general programming languages.

3.2.2   Core Ideas

Based on the above considerations, the author puts forward a new GC pattern called "Document As a System (DAS)", its core idea is under the document processing environment of MS Word or Kingsoft WPS, the business staff use the GC language to describe a process of GC to form intelligent documents that can be understood by computers, and then the intelligent document drives the background system to complete the GC (as shown in Fig. 3.2).

In the above description, business staff refer to technicians, such as planners, who have no programming ability and are not familiar with GIS software operation, but they are familiar with GC business (such as Dual-evaluation of territorial and spatial planning) and can accurately describe the GC process of each GC task, the

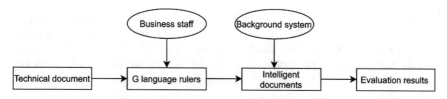

Fig. 3.2 Work flow of the new GC pattern

required geographic processing and the parameters required for each geographic processing. The intelligent document is MS word or Kingsoft WPS document, which contains the methods, process and results of GC. From the perspective of system development, the intelligent document includes a lot of notes and program codes of GC results. The reason why it is called intelligent document is that the document is intelligent. Different from the general word document, the document can not only be understood and written by technical personnel, but also read and processed by the background system. From the perspective of programming, the background system here is the interpreter of GC language, which is responsible for interpreting intelligent documents and driving the background GIS system to perform corresponding GC operations, such as overlay analysis, buffer analysis, etc.

GC Language (G Language) is the core technology in the new GC pattern. This language is a set of GC instruction set that is easy for business staff to understand and master and can be recognized and executed by computers. It has nothing to do with specific programming language and specific GIS platform. The G language defined here is a set of natural language-like programming languages designed for GC tasks based on the idea of domain-specific languages. Ordinary business staff who do not have programming ability do not need to spend a lot of energy to master it. At the same time, the GC keywords defined by this language have nothing to do with GIS platform, which means that business staff do not need to master the operation of GIS platform (such as ArcGIS of ESRI and SuperMap Software Co., Ltd), but can also use G language to carry out complex GC. In this way, under the DAS pattern, the application of GIS has stepped from GIS platform-level application to language-level application.

See Chap. 4 for a detailed introduction to G language.

3.2.3 Technical Features

Compared with the traditional GC pattern, the new GC pattern has the following technical characteristics:

First, it is easy to implement. Because the whole operation process is carried out in the familiar MS word or Kingsoft WPS, the description, GC process and GC results of all computation models are integrated in MS word or Kingsoft WPS, which avoids the switching between different systems in the GC process. At the same time, because it is no longer necessary to understand and master various complicated operations of the background GIS platform, the difficulty of GC implementation is reduced, and more and more personnel without GIS professional training can also carry out various complicated geographic processing. From then on, GIS can be used as a convenient spatial data processing tool and accepted by the public, which plays an important role in the popularization of GIS technology.

Second, the model can be adjusted. In the new GC pattern, because all the GC processes are described in MS word or Kingsoft WPS environment with easy to learn G language, business personnel can easily adjust the model and parameters of GC without the intervention of developers to meet the needs of practical work. At the same time, business staff can also conveniently carry out various tests, such as testing the influence of different thresholds on the results and discovering the mechanism of different influence factors.

Third, the quality of the results is controllable. As mentioned earlier, the existing GC pattern does not have an effective means to check the quality of large-scale GC. In the new GC pattern, the GC process (computation model and parameter setting) is completely recorded in MS word or Kingsoft WPS due to its standardized and transparent GC process, as shown in Fig. 3.3, so that the GC results can be traced back and checked. Business staff can ensure the quality of the final GC results by analyzing the correctness of each operation step. This inspection mechanism provides solid technical support for the implementation of large-scale GC.

3.2.4 Impact on the Industry

The new GC pattern not only ensures the standardized, efficient and high-quality implementation of GC, but also has a certain impact on relevant practitioners. Figure 3.4 shows the difference between DAS pattern and traditional GC pattern, In the existing GC patterns, the business staff first interprets the "technical documents" to form an "operation manual" composed of a series of data operation steps for the data operator or system developer to use. After that, the data operator implements step-by-step data operation according to the operation manual to generate data processing results, or the system developer uses visual modeling, script programming (such as Python) or general programming language (such as C #) to develop corresponding data processing tools or systems according to the operation manual to generate data processing results through the tools or systems. However, in DAS pattern, business staff directly use G language to describe the data processing process to form "intelligent documents", and then the "intelligent documents" drive the background service system to generate processing results. Under DAS pattern, business staff no longer rely on GIS professionals for data processing operations or system development, so that they can give full play to their subjective initiative and build geographic analysis models.

Based on the above description, we can draw the following four conclusions:

Steps	Operation instruction	Input	Operation	Output	Description
1.	Filter + Euclidean distance + Reclassification	[Central city]	[Description]Filter + Euclidean distance + Reclassification [Keywords] {Operation field | Filter list} % Reclassification information KX_SelDisReClass (DJ|1% 5: <80000| 4: 80000-160000| 3: 160000-280000| 2: 280000-400000|1: >=400000)	[Central city 1] DX71_ZXCS1	Calculated by 80km / h 1|2|3.5|5
2.	Filter + Euclidean distance + Reclassification	[Central city]	[Description]Filter + Euclidean distance + Reclassification [Keywords] {Operation field | Filter list} % Reclassification information KX_SelDisReClass (DJ|1% 5: <40000| 4: 40000-120000| 3: 120000-200000| 2: 200000-320000|1: >=320000)	[Central city 2] DX71_ZXCS2	Calculated by 80km / h 0.5|1.5|2.5|4
3.	Filter + Euclidean distance + Reclassification	[Central city]	[Description]Filter+ Euclidean distance + Reclassification [Keywords] {Operation field | Filter list} % Reclassification information KX_SelDisReClass (DJ|1% 5: <40000| 4: 40000-120000| 3: 120000-160000| 2: 160000-240000|1: >=240000)	[Central city 3] DX71_ZXCS3	Calculated by 80km / h 0.5|1.5|2.0|3
4.	Raster calculation	[Central city 1] [Central city2] [Central city3]	[Description]Raster calculation [Keywords]Algebraic or logical expression KX_RasCalculator ([R1] +[R2] +[R3])	[Location advantage 1] DX71_QWYSD1	
5.	Reclassification	[Location advantage 1]	[Description]Raster reclassification (equal distance), [Keywords]N1|N2|···|Nn KX_Reclass(1|2|3|4|5)	[Location advantage2] DX71_QWYSD2	
6.	Clip	[Location advantage 2] [Range layer]	[Description]Clip [Keywords]Clipped layer, Clip layer KX_Clip	[Location advantage] DX71_QWYSD	

Fig. 3.3   Example of Dual-evaluation location advantage degree calculation

First, the impact on data operators. Because there is no need for data operators to repeat their work in DAS pattern, this not only reduces the cost of GC, but also eliminates the influence of human errors, thus ensuring the quality and efficiency of GC. Therefore, the promotion of DAS pattern will have a certain impact on professionals who are only familiar with GIS operation, which also urges GIS education to make great efforts to improve students' geographic analysis ability, instead of just letting them master the operation of several GIS software. Because the operation of GIS software is not involved in DAS pattern, users only need to master the keywords of G language, and these keywords have nothing to do with specific GIS software, that is to say, no matter whether the background support system is ArcGIS or SuperMap, the unified keywords are used in G language.

Second, the impact on developers. Since system developers are no longer required to intervene in DAS pattern, business staff can adjust or develop geographic analysis

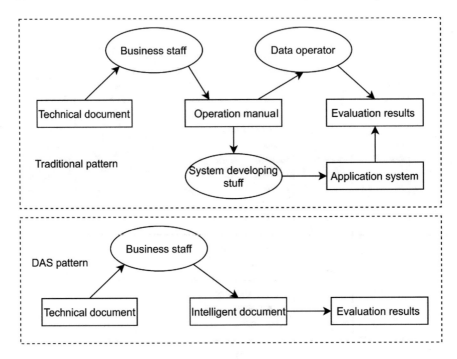

Fig. 3.4 Conventional pattern versus DAS pattern

models according to business needs. This will undoubtedly be a challenge to GIS system development practitioners, which is a problem worth thinking about by relevant personnel. After all, with the lowering of the application threshold of GIS, the future GIS, just like the current AutoCAD and MS Word, does not need to spend a lot of energy to study specially.

Third, the impact on business staff. Because DAS pattern reduces the application threshold of GIS and the development threshold of GIS analysis system, end users (such as planners) can construct their own data analysis and data processing systems according to their actual work needs. This will stimulate the creativity of countless end users and improve the efficiency and quality of data processing and data analysis. This is of great significance to explore the potential of spatio-temporal big data in the era of big data. At the same time, due to the lowering of GIS application threshold, more people can enter the GIS field, thus promoting the development of GIS industry.

Fourth, the impact on GIS education. In DAS pattern, the application pattern of GIS rises from platform-level application to language-level application. In this case, students do not have to spend too much energy to learn the operation of various GIS platform software, while teachers need to transfer their energy to how to cultivate students' geographical thinking ability, that is, to establish geographical analysis models with spatial analysis methods to solve geographical problems.

3.3   Key Technologies of the New GC Pattern

DAS pattern is a brand-new GIS application pattern, which transforms the traditional platform-level application pattern into the language-level application pattern, making the implementation of GC tasks more systematic, scientific and standardized. At the same time, for business staff without programming ability, the construction of data processing and data analysis systems is no longer a distant dream. DAS pattern involves many technical problems, such as how to construct G language with natural language features, how to edit, debug and run G language codes, how G language codes drive background GIS software, etc. This can be summed up in four aspects, namely G language, intelligent document technology, G language interpreter and G language integrated development environment, as shown in Fig. 3.5.

1. **GC language (G language)**

G language is the core technology of the new GC pattern, and the other three technologies are closely related to G language. G language mainly involves the terminology, grammar rules and keyword setting of G language. See Chap. 4 for a detailed introduction of this part.

2. **G language interpreter**

G language interpreter, or background service system, parses intelligent documents according to the grammar rules of G language, extracts keywords and computation parameters of GC, and calls the underlying GIS secondary development library to realize GC. G language interpreter is the core of the whole DAS pattern, involving the conversion from natural language to programming language, the analysis of intelligent documents and the support of G language for different GIS platforms. See Sect. 5.2 for a detailed introduction of this part.

Fig. 3.5 Key technologies of the new GC pattern

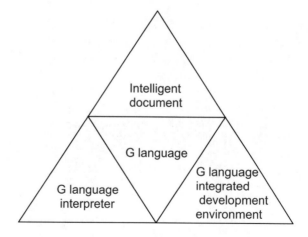

3. Intelligent documents

Intelligent documents refer to the technical documents that are normalized by business staff through G language as intelligent documents that can be understood and executed by G language interpreters. From the perspective of programming, intelligent documents can also be understood as program codes written by G language. These program codes need to be analyzed by G language interpreter and a series of GCs are performed by calling the spatial analysis function of GIS software. See Sect. 5.4 for a detailed introduction of this part.

4. G language integrated development environment

The G language integrated development environment, or G language IDE, mainly includes the editing of intelligent documents in MS Word or Kingsoft WPS, the debugging of intelligent documents and the execution of GC tasks. See Sect. 5.5 for a detailed introduction of this part.

Chapter 4
GeoComputation Language

4.1   G Language Overview

4.1.1   Raising of Questions

In DAS pattern, GC language plays a very critical role. In the existing GC pattern, visual modeling, script programming or system development are needed to realize the automatic processing of geographic data. However, due to the limitation of visual modeling and the lack of programming ability of most people, various difficulties have arisen in current GC. An analysis of the deep-seated reasons will reveal that, although the programming language that drives computers has developed from assembly language to advanced language, enabling more people to master programming technology, it is not easy to truly master a programming language to solve problems in learning and work due to various rules and the use of complex structures of programming language.

For a long time, people have been expecting to create a programming language like natural language that is easy for ordinary people to master. However, after years of efforts, people have found that it is difficult to describe what programs need to do in natural language due to the inaccuracy of natural language. For example, it is extremely inconvenient to express a mathematical expression in natural language. Therefore, it is difficult to completely replace programming language with natural language.

Although there is not a suitable programming language to meet the needs of ordinary users, However, for specific fields, such as GC project, because its GC process is basically a collection of a series of GC instructions, it is possible for us to get rid of the shackles of traditional programming languages and construct a concise, easy to understand and master domain-specific language, so that more non-programmers can use it conveniently and freely.

In view of the above situation, the author creatively puts forward the GC language (G language) in DAS pattern. The core of this language is to use the Keywords commonly used in GC (such as slope analysis, buffer analysis, network analysis,

© Surveying and Mapping Press 2021
W. Zhou, *A New GeoComputation Pattern and Its Application in Dual-Evaluation*,
https://doi.org/10.1007/978-981-33-6432-5_4

vector raster transformation, thematic mapping, statistical analysis) to describe various geographic analysis models or the process of spatial data processing. The recognition and execution of keywords can be accomplished by G language interpreter, which is based on GIS platform (such as ArcGIS, SuperMap, MapGIS and even open source GIS) and implemented by programming language (such as VB.net and Java).

4.1.2   G Language Design Idea

1. For business users who do not have programming capabilities

Generally, the target group of programming language is programmers (using C # or VB.net) or business staff with certain programming ability (using Python or R language), while the target group of G language is most business staff who do not have programming ability. For these users, if they are familiar with their own business and have a basic understanding of GIS, they can apply G language to solve GC problems.

2. Easy to learn

Since the user group of G language is business staff who do not have programming ability, in order to reduce the difficulty for them to learn G language, the grammar rules of G language should be concise, clear, easy to learn and easy to remember, so that they can understand and master it quickly.

3. Readability

G language should provide enough information to ensure that business staff can fully understand the meaning of GC tasks used or processed from methods, models to parameters, to facilitate knowledge sharing and dissemination.

4. Process traceability

In order to ensure the quality of GC, G language needs to provide a series of means to ensure the verifiability and traceability of GC process.

5. Tabular computation

In order to reduce the difficulty for ordinary business staff to use G language, G language formally uses the tables in Word documents to express the GC process. In other words, using G language to program only requires people to fill in forms according to the simple rules of G language.

6. Platform independence

G language should be a description language of GC process across GIS platform. Although specific GC is realized through existing GIS platform, the keywords used in GIS language have nothing to do with specific GIS platform. Any GIS platform can use G language for GC if the keywords set by G language are realized. In other words, for different GIS platforms (such as ArcGIS of ESRI or SuperMap Software Co., Ltd), if the interpreter of ArcGIS version or SuperMap version of G language is realized by using the secondary development library provided by these platforms, various GCs can be completed by using G language. In this way, for ordinary business staff, if they want to carry out GC, they only need to master G language without mastering the specific operation details of GIS platform. This will undoubtedly reduce the cost and application threshold of learning GIS.

4.1.3   Basic Terminology of G Language

Because G language is primarily a domain-specific language for end users, the user of the language is the end user. However, end users usually have no experience in writing programs, and terms in traditional programming languages, such as processes, functions, statements, etc., are difficult for them to understand. Therefore, concepts or terms oriented to business fields and easy to understand, such as GC project, GC task and GC item, are adopted in G language to facilitate communication and exchange between business staff.

(1)   GC project

Refers to a complex computation project involving geo-spatial analysis and research, such as Dual-evaluation, various environmental evaluations, etc. of territorial and spatial planning. From the perspective of programming, a GC project in G language is equivalent to application program projects in traditional programming language.

(2)   GC task

For a GC project, due to the large number of contents involved, a GC project is usually divided into several GC tasks according to its specialty or processing and analysis correlation, so that the GC project can be carried out reasonably and orderly. For example, for the Dual-evaluation of territorial and spatial planning, it can be divided into more than 20 GC tasks such as land resources evaluation and water resources evaluation. The concept of GC tasks corresponds to procedures or functions in traditional programming languages.

(3)   GC task group

In order to facilitate the management of GC tasks, GC tasks can be divided into several GC task groups. For example, in the Dual-evaluation of territorial and spatial planning, more than 20 GC tasks are divided into multiple groups such as individual evaluation, integrated evaluation, comprehensive evaluation, comparative analysis, etc. The concept of GC task group corresponds to the module composed of functions in traditional programming languages.

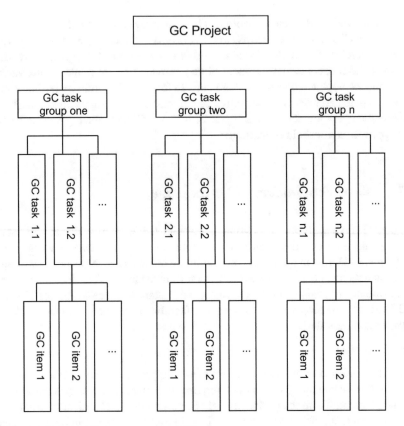

Fig. 4.1 The relationship between terms of G language

(4) GC item

For a GC task, it is composed of several GC items. Each GC item can independently complete a spatial analysis work, such as buffer analysis, overlay analysis, vector data rasterization, etc. From the point of view of programming, the GC items in G language are equivalent to the execution statements in traditional programming languages.

Figure 4.1 shows the relationship between the terms of G language.

4.2 Grammatical Rules for the G Language

The grammar of programming language is the core of programming language. Due to the complexity of the grammar of traditional programming language, only a small number of people can master it skillfully through long-term and systematic training.

For this reason, the basic principle of G language design is that grammar is as concise as possible, and the expression is simple, easy to learn and master.

Based on the above basic principles and considering that G language is a domain-specific language rather than a general programming language, the following simple syntax rules for G language are determined.

< G language > → < 1 Basic parameter table >|< 1 GC task registry >|< n GC tasks>

That is, the grammar of G language is composed of a [Basic parameter table], a [GC task registry] and a plurality of [GC tasks] filling rules.

4.2.1   Basic Parameter Table

The contents of the [Basic parameter table] in G language are equivalent to the global variables in general programming languages and are used for the information needed by relevant GC in the whole GC project. The structure of the [Basic parameter table] is as follows.

Basic parameter table = {Basic parameter item | Basic parameter value| Content description}

Note: A [Basic parameter table] consists of three parts: basic parameters, basic parameter values and content description. At present, it mainly sets 6 global quantities such as basic workspace, range layer, grid size, statistical layer, thematic map template and questionable point map template, as shown in Table 4.1. Of course, with the deepening of the research and application of G language, other global quantities can be added.

[Basic workspace] refers to the storage directory of spatial data involved in a GC project organized according to a certain mode. Figure 4.2 is an example of the basic workspace directory structure of a Dual-evaluation GC project. The [Basic workspace] can take either absolute path or relative path, and the parent path when the relative path is taken is the path where the intelligent document is located.

[Range layer] is used to set up the working range, coordinate system and clip output map of a GC project. In the Dual-evaluation GC project, the coordinate system requires the 2000 National Geodetic Coordinate System (CGCS2000) and Gauss-Kruger Projection. The [Range layer] can adopt either a relative path or an absolute path. When the relative path is adopted, the parent path is the path referred to by the [Basic workspace].

Table 4.1 Contents of [Basic parameter table]

Sequence No.	Basic parameter item	Basic parameter value	Content description
1.	[Basic workspace]	D:\QHSPJ2019P1\T20190615	Basic working directory
2.	[Range layer]	BaseMap/MapStat.shp	For setting working range, coordinate system and cropped output map
3.	[Grid size]	20	For setting the size of the pixel in meters
4.	[Statistical layer]	BaseMap/MapStat.shp	For partition statistics
5.	[Thematic map template]	Atlas1.mxd	For making thematic maps
6.	[Questionable point map template]	ZPoint.mxd	For making questionable point map

Fig. 4.2 Basic workspace example of Dual-evaluation

[Grid size] refers to the grid cell size used for raster calculation in a GC project. Different GC projects need to be set according to the requirements of actual tasks. For example, in the Dual-evaluation GC project, for the provincial (regional) Dual-evaluation, the calculation accuracy adopts a 50 m × 50 m grid, and the mountain hills or small areas can be increased to 25 m × 25 m or 30 m × 30 m. The Dual-evaluation of cities and counties adopts the calculation accuracy of 25 m × 25 m or 30 m × 30 m.

[Statistical layer] is the statistical unit layer used to generate statistical tables. The layer defined here is a global quantity that can be directly used for layer statistics in all GC tasks. However, other layers can also be specified as statistical layers for specific use (as shown in Fig. 4.3). [Statistical layer] can adopt either relative path or absolute path.

[Thematic map template] is used to output various thematic maps in GC project, in which commonly used layer styles are preset. Schedule 1 gives a list of specific styles. Please refer to Fig. 4.4 for the setting of thematic map style for specific use. [Thematic map template] can also adopt relative path and absolute path.

[Map template] is used to output various related evaluation factor maps of questionable areas in GC project (as shown in Fig. 4.5), in which the layer styles of

| 16. | Generate [the statistical table of urban construction conditions] | [Statistical layer] [Urban construction conditions] | [Description] Statistical analysis
[Keywords] Statistical layer, counted layer.
Display field \| 5,4,3,2,1 \| table serial number
KX_Statistic (Name\| 5,4,3,2,1) | | 4 |

Fig. 4.3 Sample application of [statistical layer] information

| 14. | Make [the thematic map of urban construction conditions] | Urban construction conditions | [Description] Make thematic map
[Keywords] Style \| Background style 1,
Background style 2 \| Resolution
KX_ Mapping(C3\|Boundary\|200) | [the thematic map of urban construction conditions]
DYS1_CZJS.emf |

Fig. 4.4 Layer style application example

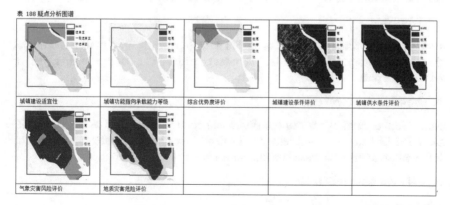

Fig. 4.5 Example of [Map template] style application

various maps are preset. The [Map template] can also adopt both relative paths and absolute paths.

4.2.2   GC Task Registry

For a GC task defined in a GC project, the background interpreter is required to identify and execute it, which requires registering each GC task that has been written in the [GC task registry] and filling in the relevant information of each GC task. The structure of the [GC task registry] is as follows:

GC task registry = {Task ID |Calculate or not| GC workspace |Input and control table| Mission description}.

1.   Task ID

Refers to the unique ID of a GC task, which usually consists of two parts, namely, GC task category name + GC task name. For example, for the land resource evaluation in Dual-evaluation, the evaluation belongs to individual evaluation category, so the task ID of the GC task is: Individual evaluation_land resource.

2.   Calculate or not

It is used for GC control in full-automatic computation of GC project, including two states of "participating in computation" and "not participating in computation". When "participating in computation" is represented by "Y", otherwise it is empty. When automatic computation is performed, the background interpretation system can select whether to perform the GC task according to the information.

3.   Workspace

Used to specify the storage path of spatial data (input data, process data and output data) participating in the GC task. The storage path can be either a relative path or an absolute path.

4.   Input and control table

Refers to the serial number of the [Input and control table] used to describe a GC task. The G language interpreter automatically locates to the table and reads the information in the table based on this information.

5.   Mission description

Refers to a brief description of GC tasks.

Table 4.2 is an example of a [GC task registration information table].

Table 4.2   Example of [GC task registration information table]

Sequence No.	Task ID	Calculate or not	Work space	Input and control table	Task description
1.	*Individual evaluation_ecology*	*Y*	*DX/DX1*	*Table 6*	*Ecological evaluation*
2.	Individual evaluation_land resource	Y	DX/DX2	Table 12	Land resource evaluation
3.	Individual evaluation_water resource	Y	DX/DX3	Table 18	Water resources evaluation
4.	Individual evaluation_climate	Y	DX/DX4	Table 25	Climate evaluation
5.	Individual evaluation_environment	Y	DX/DX5	Table 30	Environment evaluation
6.	*Integrated evaluation_ecological protection*		*JC/JC1*	*Table 62*	*Ecological protection evaluation*
7.	Integrated evaluation_agriculture production		JC/JC2	Table 66	Integrated evaluation of agricultural production
8.	Integrated evaluation_urban construction		JC/JC3	Table 70	Integrated evaluation of urban construction

4.2.3   GC Task

Normalized description of a GC task is the core content of G language. A GC task includes four parts: evaluation method, input and control, GC process and thematic map-statistical table.

> < GC task > → {< Description of evaluation method >| < Input and control table >
> |<GC process table >| < Thematic map and statistical table >}

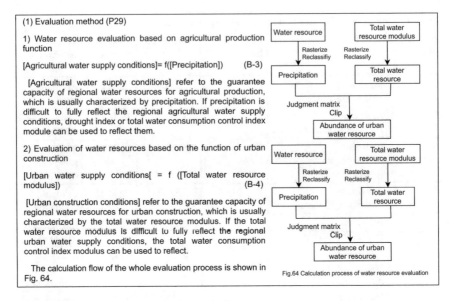

Fig. 64 Calculation process of water resource evaluation

Fig. 4.6 Example of GC task evaluation method

1. Description of evaluation method

It is a description of the mathematical model used by the GC task so that the business staff can understand the subsequent GC task. This part belongs to the annotation content, and the G language interpreter does not do any processing on this part of the content. This part of the content can be described by graphic method. Figure 4.6 is a brief description example of water resources evaluation method in the Dual-evaluation GC project.

2. Input and control table

The structure of the Input and control table is as follows:

Input and control table = Layer name |Physical layer| {Reference page} |{Description}| GC control

The contents in curly brackets are optional.

(1) Layer name

It is used to describe the Chinese name of the input layer, which is convenient for business staff to communicate.

(2) Physical layer

The physical save name for the layer for actual data processing.

(3) Reference page

Refers to the description page of the layer data in the technical documents (such as *Technical Guidelines*), which is convenient for business staff to understand the connotation of the layer data.

(4) Description

It is used to explain the values of the main attributes of the layer and their meanings.

(5) Computation control

It is used to control the actual GC process when executing GC tasks and is convenient for users to debug the GC task. The specific computation control representation method is as follows:

(a) No contents, indicating that all geographic task items are computed when performing GC tasks;
(b) Use "," to divide numbers, such as 1, 3 and 8, to indicate the first, third and eighth steps of computation; dividing numbers with ":" or "-", such as 1: 4 or 1-4, indicates 1 to 4 steps of computation, and :4 or -4 indicates 1 to 4 steps of computation; 4: or 4- means to compute 4 to the last step.

Table 4.3 is an example of [Input and control table] in the Dual-evaluation land resources evaluation GC task.

3. GC process table

This part is the main body of the GC task. The structure of the table is as follows:

GC process table = {Step number|{Operation description}| Input| Operation| Output |{Description}}

(1) Step number

For the row serial number, the content referred to by [GC control] in [Input and control table] refers to the row serial number here.

Table 4.3 Example of [Input and control table] for land resources evaluation

Sequence No.	Layer name	Physical layer	Reference page	Description
1.	[Height]	DX2_GC2	13	City and county with grid size of 20 m * 20 m or 30 m * 30 m
2.	[Soil texture]	DX2_TRZD2	14	Percentage of silt content in soil
[GC control]				
1–3				

(2) Operation description

It's a brief description of the action taken on the GC item. When running the GC task, this information will be displayed in the GC Window, prompting the user for the current operation. In order to reduce the amount of user input, "@" can be used to represent the information of the input part and "#" can represent the information of the output part. When the "GC Window" is displayed, the system will automatically replace the corresponding content.

(3) Input

In order to enter the [Layer name] of the layer, according to the different [keywords] used, one or more layers can be represented by [R1], [R2], …, [Rn] respectively in the [keyword] parameter description part. Each layer is divided by carriage return, and one line represents one layer. Layers that are usually listed in the [Input and control table] are represented here in red font.

(4) Operation

It consists of [Explanation] and [Keywords]. The [Description] section includes a description of the type of operation and additional instructions (this section is optional). The explanation part is only for the convenience of business staff to understand the operation, and the G language interpreter will not do any processing on this part. [Keywords] is the core content of the whole G language, it is also the key to realize document intelligence. This part includes keyword marker and keyword. The keyword marker is [keywords]. The G language interpreter will recognize the marker and the next line of the marker is recognized as "keyword". The keyword marker can explain the meaning of control parameters in the keyword. The content in the [keywords] line is a keyword expression, which is usually composed of G language keyword identification and control parameters. The following is an example of the [keywords] line.

KX_Reclass (1: <1500| 2: 1500–4000| 3: 4000–5800 |4: 5800–7600 |5: ≥7600)

Among them, KX_Reclass is the keyword marker, and "1: <1500 2: 1500–4000 3: 4000–5800 4: 5800–7600 5: ≥7600" is the control parameter.

Individual G language keywords have no control parameters, such as KX_ExtractByMask.

See sect. 4.3 for the specific introduction of G language keywords. Figure 4.7 is an example of the GC process of a GC task.

(5) Output

It is the output content of keywords, mainly including vector or raster layers, thematic maps, statistical tables, and statistical charts. For layer objects, thematic maps and

Steps	Operation instruction	Input	Operation	Output	Description
1.	Active accumulated temperature	[Meteorology]	[Description] Interpolation [Keywords] Interpolation field | Interpolation method (IDW, KRI | Resolution 20 KX_interpolation (HDJW|IDW)	[Photothermal conditions 1] DX4_GRTJ1	[Active accumulated temperature]
2.	DEM modification	[Photothermal conditions 1] [Height]	[Description]Raster calculator [Keywors] KX_RasCalculator([R1]-[R2]/100*0.6)	[Photothermal conditions 2] DX4_GRTJ2	
3.	Reclassification of [photothermal condition 2]	[Photothermal conditions 2]	[Description] Reclassification (numerical range) [Keywords] KX_Reclass(1: <1500| 2: 1500-4000| 3:4000-5800 |4:5800-7600 |5: >=7600)	[Photothermal conditions 3] DX4_GRTJ3	
4.	@Clipped by ranger layer	[Photothermal conditions 3] [Range layer]	[Description]Clip [Keywords]Clipped layer, Clip layer KX_ExtractByMask	[Photothermal conditions] DX4_GRTJ	5(Excellent) 4(good) 3(Common) 2(Poor) 1(Bad)

Fig. 4.7 Example of GC process

statistical maps, each output has a Chinese name corresponding to a physical name. Among them, the Chinese name can be quoted in the input part in the subsequent operation. At the same time, the Chinese name is also the interpretation of the physical name, which is convenient for people to understand the GC pattern. The physical name is the physical storage location of the processing results, using the relative path, and its parent path is the path where the GC task is located (set in the [GC task registry]). For statistics and picture insertion operations, the output content is the table serial number output by statistical tables and thematic maps.

(6) Description

The description part is an additional description of this GC item, which can be used flexibly and is usually an explanation of the output content.

4. **Thematic map and statistical table**

This part is the performance or visualization of the results of the GC task. The thematic maps and statistical tables generated in the GC process can be output in the designated tables, so that business staff can observe and analyze the GC results in time, or directly use the chart results of this part in the subsequent report preparation. This part includes thematic charts and statistical tables.

(1) Thematic chart

Thematic charts are tables used to display thematic charts. The contents of the tables are generated by KX_Mapping keyword and filled into the specified tables

| 11. | Make thematic map of farming conditions | [Farming conditions] | [Description] Make thematic map
[Keywords] Replace list \| Background list \| resolution \| template \|Positioning statement
KX_Mapping (C2\|Boundary \|200) | [Thematic map of farming conditions]
DX2_NYGZTJ.emf | |
| 12. | Insert thematic maps | [Thematic map of farming conditions] | [Description]Insert thematic map
[Keywords]Picture height
KX_InsertPic (12) | | 1 |

XX市资源环境承载能力和国土空间开发适宜性试评价
农业耕作条件专题图

Fig. 4.8 Example of output pattern of single thematic map

by KX_InsertPic keyword. Thematic charts can adopt single chart pattern (as shown in Fig. 4.8) or multi chart pattern (as shown in Fig. 4.9).

(2)   Statistical table

The statistical table is used to display the statistical data of thematic maps. The contents in the table are the results of statistical analysis of thematic layers according to specified statistical layers through KX_Statistical keyword. The statistical tables can also adopt single-graph statistical pattern (as shown in Fig. 4.10) or multi graph statistical pattern (as shown in Fig. 4.11).

4.3   Keywords in G Language

Each language has its own keywords or reserved words. For example, C language defined by ANSI standard has 32 keywords, such as break, continue, do, else, double, for, return, short, void, while, etc. Similarly, G language also includes a series of keywords, which are used in the [Keyword] of the [GC process table].

Different from the keywords in general programming languages, the keywords in G language are mainly used for spatial analysis. Because there are many operations for spatial analysis, such as hundreds in ArcGIS, too many keywords are a burden for users to learn, while too few keywords cannot meet the needs of practical applications.

Steps	Operation instruction	Input	Operation	Output	Description
1.	Clip @	[Suitability of agricultural production] [Range layer]	[Description] Clip [Keywords] Clipped layer, Clip layer KX_ExtractByMask	[Suitability of agricultural production Z] DB3_NYSCSYXZ	
2.	Produce [city and county level agricultural production suitability evaluation map]	[Suitability of agricultural production Z]	[Description]Make Thematic map [Keywords]Replace list| Background list| Resolution |Template |Positioning statement KX_Mapping(C2|Boundary|200)	[city and county level agricultural production suitability evaluation map] DB3_ SXDB_3.EMF	
3.	Clip @	[Suitability of agricultural production-Province] [Range layer]	[Description]Clip [Keywords] Clipped layer, extract layer KX_ExtractByMask	[Suitability of agricultural production-Province Z] DB3_NYSCSYXSZ	
4.	Make [provincial agricultural production suitability evaluation map]	[Suitability of agricultural production-Province Z]	[Description] Make Thematic map [Keywords] Replace list| Background list| Resolution |Template |Positioning statement KX_Mapping(C2|Boundary|200)	[provincial agricultural production suitability evaluation map] DB3_ SXDB _4.EMF	
5.	Insert map	[city and county level agricultural production suitability evaluation map] [provincial agricultural production suitability evaluation map]	[Description Make Thematic map [Keywords]Picture height | Delete words / blank (not marked) KX_InsertPic (11)		1

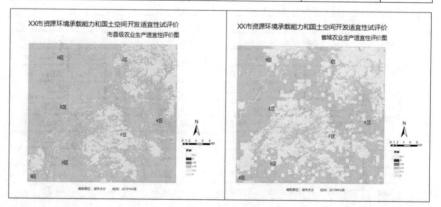

Fig. 4.9 Example of output pattern of multi thematic map

13.	Produce [Statistical table of agricultural cultivation conditions]	[Statistical layer] [Agricultural cultivation conditions]	[Description]Statistical analysis [Keywords]Statistical fields| List of Statistics| Serial number list KX_Statistic (Name|5,4,3,2,1)		2

Region		Higher		High		Medium		Low		Lower	
		Area	Ratio	Area	Ratio	Area	Ratio	Area	Ratio	Area	Ratio
	B	1.02	8.1	4.70	37.4	5.16	41.0	1.56	12.4	0.14	1.1
	D	0.87	0.8	24.63	22.7	53.59	49.5	23.78	21.9	5.47	5.0
	E	93.29	20.5	121.73	26.8	106.92	23.5	91.38	20.1	41.48	9.1
*	F	21.67	5.8	79.64	21.2	127.91	34.0	102.22	27.2	44.78	11.9
*	H	51.19	27.4	67.79	36.3	40.68	21.8	20.25	10.8	6.81	3.6
City	J	13.91	9.1	28.83	18.9	46.01	30.2	44.28	29.0	19.34	12.7
	K	109.10	26.2	116.17	27.8	78.32	18.8	74.23	17.8	39.45	9.5
	Total	291.04	17.0	443.48	26.0	458.60	26.8	357.72	20.9	157.46	9.2

Fig. 4.10   Example of statistical pattern of single-graph thematic map

6.	Statistical analysis (area)	[Statistical layer] [Suitability of agricultural production Z] [Suitability of agricultural production-Province Z]	[Description]Statistical analysis [Keywords]Display fields|3,2,1 KX_Statistic (Name|3,2,1|0,1,7,3,9,5,11)		2

Region		Suitable		General		Unsuitable	
		City	Province	City	Province	City	Province
	B	11.75	9.74	0.00	2.51	0.83	0.33
	D	102.92	74.26	0.00	31.47	5.42	2.61
	E	408.46	274.64	0.00	160.36	46.34	19.79
*	F	259.27	210.40	0.00	147.91	116.95	17.91
*	H	184.54	141.07	0.00	38.55	2.19	7.11
City	J	104.02	58.89	0.00	82.58	48.35	10.90
	K	310.18	240.52	0.00	152.56	107.09	24.19
	Total	1381.14	1009.52	0.00	615.94	327.17	82.83

Fig. 4.11   Example of statistical pattern of multi-graph thematic map

Therefore, the reasonable setting of keywords is the key to whether G language can be accepted by end users.

4.3.1 Keyword Selection Principle

The setting of keywords should follow the following principles:

1. **Systematic**

The keywords of G language should form a system for GC applications, so that the minimum keywords can be used to meet the needs of applications to the greatest extent.

2. **Encapsulation**

Keywords are not simple correspondence of GIS functions, and the convenience and integrity of functions should be fully considered. This can draw lessons from the idea of component technology in software development. The core idea of component technology is to encapsulate some details that people care about but are not convenient for users to operate directly. For example, in GIS operation, adding attribute fields to vector layer is a common operation, but it is a lower-level operation. In G language, there will be no keywords corresponding to this operation alone, but this operation will be encapsulated.

3. **Scenario**

When designing keywords, the needs of application scenarios should be fully considered and the functions of each keyword should be reasonably determined. For example, for the keyword KX_SelDisReClass, its function is to filter vector data first, then perform Euclidean distance operation, and finally perform raster data reclassification operation. KX_SelDisReClass is a common application scenario for GC.

4. **Scalability**

G language should be a continuously developed and perfected language. With the continuous expansion of the application field of G language, the existing keyword library cannot cover the application needs of other fields. Therefore, it is necessary to have a keyword extension mechanism, which allows developers to make necessary extensions to the keyword library, thus forming keyword libraries for different application fields and enabling G language to form a good ecology.

4.3.2 Expression of A Keyword

The expression of a keyword includes two parts: keyword ID and control parameters.

Keyword = {Keyword ID |Control parameters}

1. Keyword ID

The keyword ID represents the operation carried out, and its naming is as consistent as possible with the traditional naming, usually expressed in English. Considering the convenience of users, keyword ID is not case sensitive. For example, KX_RasCalculator and KX_RASCALCULATOR both represent the same operation.

2. Control parameter

The parameters corresponding to a keyword include three parts, besides the input parameters and output parameters in the [GC process table], there are also the control parameters discussed here. The contents of the input parameters are generally vector or raster layer objects, and the output parameters are generally vector or raster layer objects, while the control parameters are internal logic control parameters when the input layer is processed by keywords, and different keywords correspond to different control operations. As shown in Fig. 4.12, KX_Interpolation keyword, its control parameters are interpolation fields, interpolation methods and number of search points. The control parameter of KX_Reclass is the classification method expression, while KX_Clip does not need to call any parameters.

As mentioned earlier, keywords in G language usually include multiple functions, such as KX_SelDisReClass, which includes three functions: selection element,

Steps	Operating instructions	Input	Operation	Output	Description
1.	Spatial interpolation	[Soil pollution site]	[Description] Spatial interpolation [Keywords] Interpolation field | {Interpolation method} | {Number of search points} KX_Interpolation (TRWR)	[Soil environmental capacity 1] DX5_TRHJRL1	
2.	Reclassification	[Soil environmental capacity 1]	[Description] Reclassification (numerical range) [Keywords] KX_Reclass (3: <100|2:100-150 | 1: >=150)	[Soil environmental capacity 2] DX5_TRHJRL2	3（High）/2（Medium）/1Low）
3.	Clip	[Soil environmental capacity 2] [Range layer]	[Description]Clip [Keywords] KX_Clip	[Soil environmental capacity] DX5_TRHJRL	

Fig. 4.12 Example of keyword control parameters

Euclidean distance, and reclassification. For such keywords, the control parameters expressing each function are divided by "%" sign, such as:

KX_SelDisReclass (DJ|1% 5: <80,000 4: 80,000–160,000 3: 160,000–280,000 2: 280,000–400,000 1: ≥400,000)

Among the above control parameters, "DJ|1" is the filter parameter, and "% 5: <80,000 |4: 80,000–160,000|3: 160,000–280,000| 2: 280,000–400,000|1: ≥400,000" is the reclassification parameter.

In each function parameter, "|" is used to split parameter items, and "," is used to split list items, such as:

KX_SelectFeatureOR (DLMC| Forest land, shrub land, other forest land, artificial pasture land, other grassland)

In the above control parameters, "|" is used to split the filter fields and filter criteria, and "," is used to split the filter criteria.

In addition, for a parameter sub-item, "#" can also be used for further segmentation, and the information after "#" usually indicates the parameters required for processing the parameter, such as:

KX_Mapping (VC5 # MIN100, DBWD|Boundary|200)

In the above control parameters, "#" is used to split the layer style and the field name to be expressed.

4.3.3 Keyword Setting

According to the above principles, the functions of spatial analysis are summarized, and 38 keywords in 5 categories (as shown in Table 4.4) are obtained in combination with the actual needs of the Dual-evaluation work. These 5 categories are vector layer analysis, single raster layer analysis, multiple layer analysis, big data acquisition and output. These keywords constitute the basic keyword library of G language. With the continuous popularization of DAS pattern, its application fields will also continue to expand. At this time, in order to meet the needs of application, domain keyword

Table 4.4 G language keyword classification table

Keyword category	Keyword (label, name)	Description
Vector layer analysis class	A001, KX_Interpolation	Spatial interpolation
	A002, KX_FeaturetoRaster	Rasterization
	A003, KX_SelectFeatureOr	Feature filtering
	A004, KX_SelectFeatureOrRas	Feature filtering + Rasterization
	A005, KX_ServiceAreaExt	Accessibility of transportation network
	A006, KX_RasReclass	Rasterization + Reclassification
	A007, KX_SelDisReClass	Filtering + Euclidean distance + Reclassification
	A008, KX_Density	Conventional density calculation
	A009, KX_Density2	Kernel density calculation
	A010, KX_Reclass_Vec	Vector reclassification
	A011, KX_ModifyRaster	Raster modification
	A012, KX_EF_FieldCalculator	Factor calculation (field computation)
	A013, KX_EF_StandardField	Factor calculation (field normalization)
	A014, KX_EF_Buffer	Factor calculation (buffer)
	A015, KX_EF_Distance	Factor calculation (nearest distance)
	A016, KX_EF_Info	Factor calculation (attribute information)
Single raster layer analysis class	B001, KX_Reclass	Raster reclassification
	B002, KX_Slope	Slope calculation
	B003, KX_Undulation	Undulation calculation
	B004, KX_Aggregate_Ras	Raster aggregation
	B005, KX_GetArea_Ras	Patch area calculation
	B006, KX_GetShapeIndex_Ras	Patch index calculation
	B007, KX_FillNoData	Fill in null values
Multiple raster layer analysis class	C001, KX_RasCalculator	Raster calculation
	C002, KX_Matrix	Judgment matrix integration
	C003, KX_Matrix2	Multi-layer logic integration
	C004, KX_Max	Take the maximum value
	C005, KX_Min	Take the minimum value
	C006, KX_RasErase	Raster deduction

(continued)

Table 4.4 (continued)

Keyword category	Keyword (label, name)	Description
	C007, KX_RasIntersection	Raster intersection
Big data acquisition class	D001, KX_DB_GetPOI	Get POI
	D002, KX_ DB_GetLOI	Get LOI
	D003, KX_ DB_GetAOI	GET AOI
Output class	E001, KX_ Clip	Clip raster with vector
	E002, KX_Mapping	Make thematic map
	E003, KX_Atlas	Make thematic atlas
	E004, KX_InsertPic	Insert thematic map
	E005, KX_Statistic	Output statistical table

libraries can be built based on basic keyword libraries according to the characteristics of different fields.

4.4 Keyword Usage Instructions

This section introduces the use of 38 keywords in 5 categories in more detail. The introduction of each keyword mainly includes keyword ID, function description, input, output, control parameters, examples, and example description, etc. Among them, the control parameters are the focus of the description. This part includes [Brief description] and [Description]. The [Brief description] part is a simple description of the control parameters. The contents of this part will be used in the [Keyword] of the [GC process table] to facilitate users to understand the meaning of the control parameters. It should be noted that curly braces are used in the [Brief description] to represent the optional parts. The [Description] section is a detailed description of the control parameters.

4.4.1 Keywords for Vector Layer Analysis Class

Vector layer analysis keywords mainly refer to common keywords related to vector layer operation, mainly including spatial interpolation, vector data rasterization, network analysis and density analysis. See Table 4.5 for details.

Table 4.5 Vector layer class keywords table

Number + function	*A001 Spatial interpolation*
Keyword	**KX_Interpolation**
Function description	The inverse distance weight method or Krugin method is used for spatial interpolation
Input	Point layer
Output	Raster layer
Control parameters	**[Brief description]** Interpolation field| {Interpolation method} | {Number of search points} **[Description]** Interpolation method: IDW, KRI#{model} IDW means inverse distance weighting method, KRI means Krugin method. The model includes S-Spherical, C-Circular, E-Exponential, G-Gaussian, L-Linear interpolation method. The interpolation method is IDW by default Number of search points: 12 by default
Example	**[Description]** Spatial interpolation **[Keywords]** Interpolation field| {Interpolation method} | {Number of search points} **KX_Interpolation (Value| IDW)**
Number + function	*A002 Rasterization*
Keyword	**KX_FeaturetoRaster**
Function description	Rasterize the vector layer according to the specified field or constant
Input	Vector layer (point, polyline, or polygon)
Output	Raster layer
Control parameters	**[Brief description]** Operation field or value| {Grid size} **[Description]** Operation field or value: when rasterizing, you can specify either the field of vector layer or a constant directly, which means that the whole research range is the same value Grid size: it refers to the grid cell size, which is set in the [Basic parameter table] by default
Example	**[Description]** Vector rasterization **[Keywords]** Operation field or value| {Grid size} **KX_FeatureToRaster (JYL)**
Number + function	*A003 Feature filtering*
Keywords	**KX_SelectFeatureOr**
Function description	Select a figure or figures by logic or condition
Input	Vector layer (point, polyline, or polygon)
Output	Vector layer
Control parameters	**[Brief description]** Filter field | Filter item list **[Description]** Filter item list: each condition item is separated by ","

(continued)

Table 4.5 (continued)

Example	**[Description]** Feature filtering **[Keywords]** Filter field I Filter item list **KX_SelectFeatureOr (DLMCI Forested land, Shrubbery, Other woodland, Artificial grassland, Other grassland)**
Number + function	*A004 Feature filtering + Rasterization*
Keyword	**KX_SelectFeatureOrRas**
Function description	Firstly, the figures are filtered, and then they are rasterized according to the specified field
Input	Vector layer (point, polyline or polygon)
Output	Raster layer
Control parameters	**[Brief description]** Filter field I Filter item list I Rasterization field **[Description]** Filter list: see **kx_ SelectFeatureOr**
Example	**[Description]** Vector filtering + Rasterization **[Keywords]** Filter field I Filter item list IRasterization field **KX_SelectFeatureOrRas (DLMCI City, townI LXDWMJ)**
Number + function	*A005 Accessibility analysis of transportation network*
Keyword	**KX_ServiceAreaExt**
Function description	The coverage of service facilities is calculated based on the road network analysis function
Input	Source layer, network layer
Output	Raster layer
Control parameters	**[Brief description]** {Filter field I Filter item list} %Time unit I Time list% Reclassification information **[Description]** Filter list: see **kx_ SelectFeatureOr** Time unit: H-H our, m-minute Reclassification list: see **kx_ Reclass**
Example	**[Description]** Accessibility analysis of transportation network **[Keywords]** {Filter field I Filter item list} %Time unit I Time list% Reclassification information **KX_ServiceAreaExt(%MI30,60,90,120%5,4,3,2)**
Number + function	*A006 Rasterization + Reclassification*
Keyword	**KX_RasReclass**
Function description	Firstly, the rasterization operation is performed, and then the reclassification is carried out
Input	Vector layer (polygon)
Output	Raster layer
Control parameters	**[Brief description]** Operation field IDefault value % Reclassification information **[Description]** Default value: replaces the value of NoData when rasterized Reclassification information: see **KX_ Reclass**

(continued)

Table 4.5 (continued)

Example	**[Description]** Rasterization + Reclassification **[Keywords]** Operation field I Default value % Reclassification information **KX_S RasReclass (AREA # 0 % 1I2I3I4I5)**
Example description	The AREA field is rasterized and then reclassified at intervals
Number + function	*A007 Filtering + Euclidean distance + Reclassification*
Keyword	**KX_SelDisReClass**
Function description	Firstly, the elements are filtered, then the Euclidean distance analysis is carried out, and finally the reclassification is carried out. When the input layer includes cost layer, cost distance analysis is used instead of Euclidean distance analysis
Input	Vector layer (point, polyline or polygon)
Output	Raster layer
Control parameters	**[Brief description]** {Operation field IFilter list} % Reclassification information **[Description]** Operation field: Euclidean distance operation is performed by default Filter list: see **kx_ SelectFeatureOr** Reclassification information: see **KX_ Reclass**
Example 1	**[Description]** Filtering + Euclidean distance + Reclassification [Keywords] {Operation field IFilter list} % Reclassification information **KX_SelDisReClass (DJI1 % 5: <80,000I 4: 80,000–160,000I 3: 160,000–280,000I 2: 280,000–400,000I1: ≥400,000)**
Example 1 description	Firstly, the features are filtered by DJ = 1, then Euclidean distance operation is performed, and finally rasterized
Example 2	**[Description]** Filtering + Euclidean distance + Reclassification **[Keywords]** {Operation field I Filter list} % Reclassification information **KX_SelDisReclass (AREA#0 % 1I2I3I4I5)**
Example 2 description	The AREA field was rasterized and then reclassified with interval method
Number + function	*A008 Conventional density calculation*
Keyword	**KX_Density**
Function description	According to the conventional density computation method, compute the density of statistical units, such as building density and plot ratio
Input	Statistical unit layer (polygon), statistical layer (point, polyline, polygon)
Output	Vector layer
Control parameters	**[Brief description]** Density field I Key field I {Multiplication field} **[Description]** Key field: It is the unique value field in the layer to be counted Multiplication field: a field used to multiply with the statistic and used in calculations such as plot ratio

(continued)

Table 4.5 (continued)

Example	[Description] Conventional density calculation [Keywords] Density field | Key field | {Multiplication field} **KX_DENSITY (Density| OBJECTID)**
Number + function	*A009 Kernel density calculation*
Keyword	**KX_Density2**
Function description	The Kernel density is calculated by the method of Kernel density
Input	Point or polyline layer
Output	Raster layer
Control parameters	**[Brief description]** {Multiplication field} |Radius **[Description]** Multiplication field: the weight field involved in calculation Radius: search radius, in meters
Example	[Description] Kernel density calculation [Keywords] Multiplication field| Radius **KX_DENSITY2 (1500)**
Number + function	*A010 Vector reclassification*
Keyword	**KX_Reclass_Vec**
Function description	The specified fields are reclassified and converted to raster data as needed
Input	Vector layer
Output	Vector layer or raster layer
Control parameters	**[Brief description]** Operation field, Target field% Reclassification information **[Description]** Target field: when the output is in S hp format, the target field is the field name after reclassification; When exported to a raster layer, this information is the default value for rasterization Reclassification information: see **KX_ Reclass**
Example	[Description] Vector reclassification (class modification) [Keywords] Operation field, Target field% Reclassification information **KX_Reclass_Vec (CODE, −1%1: 01010003|2: 01010004|3: 01010002|4: 01020200|5: 01010001)**
Number + function	*A011 Raster modification*
Keyword	**KX_ModifyRaster**
Function description	Replaces the value within the range of vector layer features in the raster layer with the specified field or value
Input	Raster layer, vector layer
Output	Raster layer

(continued)

Table 4.5 (continued)

Control parameters	**[Brief description]** Field name or value **[Description]** Field name or value: if it is set as a field, the value in the field is used as the correction parameter; If it is a numerical value, the number is directly used as the correction parameter
Example 1	**[Description]** Raster modification **[Keywords]** Field name or value **KX_MODIFYRASTER (0)**
Example 1 description	Replace the values within the range of vector layer elements in the raster layer with 0
Example 2	**[Description]** Raster modification **[Keywords]** Field name or value **KX_MODIFYRASTER (LV)**
Example 2 description	Replace the values in the vector layer element range in the raster layer with the values in the "LV" field in the vector layer
Number + function	*A012 Factor (field) calculation*
Keyword	**KX_EF_FieldCalculator**
Function description	Field operation
Input	Vector layer or table list
Output	Vector layer
Control parameters	**[Brief description]** Identification field # Target field ǀ Field list ǀ Calculation expression ǀ Normalization} **[Description]** Identification field: the unique value field used by each layer or table Field list: fields from each layer are divided by ",", and fields from the same layer are separated by "#". In the field list, geomeric quantities G_Area, G_Length, G_X, G_Y can be directly used to represent the area, length, X and Y coordinate of the feature respectively Calculation expression: it is an arithmetic expression, each variable is represented by [FN], and N is consistent with the order in the field list Normalization: 0-no treatment, 1-normalization, 2-reverse normalization, by default, it means no processing
Example	**[Description]** Factor (field) calculation **[Keywords]** Identification field # Target field ǀ Field list ǀ Calculation expression ǀ Normalization} **KX_EF_FieldCalculator(FID#DFieldǀAREA#LENGTH,SUMǀ[F1]*[F2]*[F3])**
Example description	The fields AREA and LENGTH are from the first input layer, SUM is from the second input layer, DField=[F1]*[F2]*[F3]
Number + function	*A013 Factor calculation (field normalization)*
Keyword	**KX_EF_StandardField**
Function description	Normalize the specified field

(continued)

Table 4.5 (continued)

Input	Vector layer
Output	Vector layer
Control parameters	**[Brief description]** Processing field| {Normalization} **[Description]** Normalization: 0-no treatment, 1-normalization, 2-reverse normalization, by default, it means no processing
Example 1	**[Description]** Factor calculation (field normalization) **[Keywords]** Processing field| {Normalization} **KX_EF_StandardField (AREA|1)**
Number + function	*A014 Factor calculation (buffer)*
Keyword	**KX_EF_Buffer**
Function description	Firstly, the buffer is generated, and then the statistical items in the buffer are counted. Finally, normalization is performed according to the needs
Input	Vector layer, statistical raster layer
Output	Vector layer
Control parameters	**[Brief description]** Target field # Statistical item | Normalization | Buffer distance **[Description]** Statistical items: including MIN,MAX,MEAN and SUM, which respectively represent the minimum value, maximum value, average value and sum. When they are other field names, they represent [Field] * Sum. Normalization: see **KX_EF_StandardField** Buffer distance: Form 1, Buffer distance, indicates that the output buffer includes the element itself and the buffer; Form 2, Buffer Distance 1 − Buffer Distance 2, means that the output buffer is the buffer obtained by buffer distance 2 minus the buffer obtained by buffer distance 1, and is a buffer band
Example 1	**[Description]** Factor calculation (buffer) **[Keywords]** Target field # Statistical item | Normalization | Buffer distance **KX_EF_Buffer (MIN#MIN|0|100)**
Example 1 description	Firstly, a 100 m buffer (including the feature itself) is generated, and then the minimum value in the buffer is counted and stored in the min field
Example 2	**[Description]** Factor calculation (buffer) **[Keywords]** Target field # Statistical item | Normalization | Buffer distance **KX_EF_Buffer (MAX#Max|1|100–200)**
Example 2 description	Firstly, the 100–200 m buffer is generated, and then the maximum value in the buffer is counted and stored in the max field. Finally, the normalization is performed
Example 3	**[Description]** Factor calculation (buffer) **[Keywords]** Target field # Statistical item | Normalization | Buffer distance **KX_EF_Buffer(S1#Sum|2|0)**

(continued)

Table 4.5 (continued)

Example 3 description	Firstly, the sum of raster values within the range of features (without buffer) is counted and stored in S1 field. Finally, reverse normalization is performed
Example 4	**[Description]** Factor calculation (buffer) **[Keywords]** Target field # Statistical item I Normalization I Buffer distance **KX_EF_Buffer(S2#DMI0I0–100)**
Example 4 description	Firstly, a 100 m buffer (excluding elements) is generated, then the sum of buffer raster values is counted, multiplied by DM field values, stored in S2 field, and finally reverse normalization is carried out
Number + function	*A015 Factor calculation (nearest distance)*
Keyword	**KX_EF_Distance**
Function description	Calculate the nearest distance between each element of the main vector layer and the element of the acting vector layer
Input	Main vector layer, active vector layer
Output	Vector layer
Control parameters	**[Brief description]** Target field {Normalized field} {Normalization} **[Description]** Normalization field: the target field is used as the normalization field by default Normalization: see **kx_EF_StandardField**. Normalization is not performed by default
Example	**[Description]** Factor computation (nearest distance) **[Keywords]** Target field # {Normalized field} I {Normalization} **KX_EF_Distance (DF#DF1I1)**
Example description	Calculate the minimum distance between the main vector layer elements and the action vector layer elements, store them in the DF field, and carry out normalization processing. The results are stored in DF1
Number + function	*A016 Factor calculation (attribute information)*
Keyword	**KX_EF_Info**
Function description	Output attribute information in the specified table
Input	Vector layer list
Output	Output table number or text file name
Control parameters	**[Brief description]** Identification field# {Display name} I Field list **[Description]** Identification field: the unique value field used by each layer, which is used to connect the information of each layer Display name: used to display in a table or text file. If default, the identification field is used Field list: fields from each layer are divided by ",", and fields from the same layer are separated by "#"

(continued)

Table 4.5 (continued)

Example	[Description] Factor computation (attribute information) [Keywords] Identification field I field list **KX_EF_Info (FID# Namel AREA#LENGTH#AL, MIN#MAX#MEAN)**
Example description	AREA, LENGTH, and AL (from the first input layer), MIN, MAX, and MEAN (from the second layer) will be output in specified table or text file

4.4.2 Keywords for Single Raster Layer Analysis Class

Single raster layer analysis keywords mainly refer to common keywords related to single raster layer operation, and their using methods are shown in Table 4.6.

Table 4.6 Single raster layer analysis keywords table

Number + function	*B001 Reclassification*
Keyword	**KX_Reclass**
Function description	It supports the reclassification of raster data and vector data, and supports three classification methods, namely class modify type, numerical range type and equal distance type
Input	Raster or vector layer
Output	Raster or vector layer
Control parameters	1. Class modification type **[Brief description]** N1: C1, C2, …IN2: C3, C4, …I… **[Description]** N1: New class value C1: Old class value 2. Range value type **[Brief description]** N1: <V1IN2: V1 − V2I…INn: ≥Vn **[Description]** Nn: The nth category value Vn: The nth partition value 3. Equal distance type **[Brief description]** N1IN2I…INn **[Description]** Nn: The nth category value
Example 1	[Description] Reclassification (class modification) [Keywords] N1: C1, C2, …IN2: C3, C4, …I… **KX_Reclass (3:5I2:4,3,2I1:1)**

(continued)

Table 4.6 (continued)

Example 2	**[Description]** Reclassification (numerical range) **[Keywords]** N1: <V1|N2: V1 − V2|…|Nn: ≥Vn **KX_Reclass (1: <0.25| 2: 0.25–0.5| 3: 0.5–1.0 |4: 1.0–2.0 | 5: ≥2.0)**
Example 3	**[Description]** Reclassification (equal distance) **[Keywords]** N1|N2|…|Nn **KX_Reclass (5|4|3|2|1)**
Example 4	**[Description]** Reclassification (class modification for vector layer) **[Keywords]** N1: C1, C2, …|N2: C3, C4, …|… **KX_Reclass (DLBM|0#1:0101|2:0102|3:0103)**
Number + function	*B002 Slope calculation*
Keyword	**KX_Slope**
Function description	Calculate the slope according to DEM
Input	Raster layer
Output	Raster layer
Control parameters	None
Example	**[Description]** Slope calculate **[Keywords]** **KX_Slope**
Number + function	*B003 Undulation calculation*
Keyword	KX_Undulation
Function description	The undulation is computed according to the given neighborhood range
Input	Raster layer
Output	Raster layer
Control parameters	**[Brief description]** Neighborhood size **[Description]** Neighborhood size: the size of the range in pixels for neighborhood analysis
Example	**[Description]** Undulation calculation **[Keywords]** Neighborhood size **KX_Undulation (21)**
Number + function	*B004 Raster aggregation*
Keyword	**KX_Aggregate_Ras**
Function description	Aggregate with the specified pixel distance
Input	Raster layer
Output	Raster layer
Control parameters	**[Brief description]** Aggregation distance **[Description]** Aggregation distance: the distance, in pixels, at which the aggregation operation is performed

(continued)

Table 4.6 (continued)

Example	**[Description]** Aggregation operation **[Keywords]** Aggregation distance **KX_Aggregate_Ras (2)**
Number + function	*B005 Patch area calculation*
Keyword	**KX_GetArea_Ras**
Function description	Calculate the patch area of raster layer
Input	Raster layer
Output	Raster layer
Control parameters	**[Brief description]** Unit **[Description]** Unit: unit of area, expressed in km or mu, where km stands for square kilometers, mu is an area unit used in China, and one mu is equal to 666.66 m^2
Example	**[Description]** Patch area calculation **[Keywords]** unit **KX_GetArea_Ras (km)**
Number + function	*B006 Patch index calculation*
Keyword	**KX_GetShapeIndex_Ras**
Function description	Calculate the patch shape index or compactness of the raster layer
Input	Raster layer
Output	Raster layer
Control parameters	**[Brief description]** Shape index or compactness **[Description]** Shape index or compactness, represented by zzzs and JCD, where xzzs represents shape index and JCD represents compactness
Example	**[Description]** Patch index calculation **[Keywords]** Shape index or compactness **KX_GetShapeIndex_Ras (XZZS)**
Number + function	*B007 Fill in null value*
Keyword	**KX_FillNoData**
Function description	Fill in the null value with the given value
Input	Raster layer
Output	Raster layer
Control parameters	**[Brief description]** Fill value
Example	**[Description]** Fill in null value **[Keywords]** Fill value **KX_FillNoData (1)**

4.4.3 Keywords for Multi-raster Layer Analysis Class

Multi-raster layer analysis keywords mainly refer to common keywords related to multi-raster layer operations, and their usage methods are shown in Table 4.7.

Table 4.7 Multi-raster layer analysis keywords table

Number + function	*C001 Raster calculation*
Keyword	**KX_RasCalculator**
Function description	Raster layer logical or algebraic calculation
Input	Raster layer list
Output	Raster layer
Control parameters	**[Brief description]** Algebraic or logical expressions **[Description]** Logical expression: condition 1, result 1%, condition 2, result 2%… %Other results (if none, NoData). In the conditional expression, operator "and" and "or" are supported Arithmetic expression: map algebraic expression The variable in the expression is represented by [Rn], where n is the ordinal number of the input raster layer Note: the maximum length of an expression is 4096 characters
Example 1	**[Description]** Raster calculation If [Height] > 4000, then [Land resource 2] = [Land resource 2] − 2; If 3000 < [Height] < 4000, [Land resource 2] = [Land resource 2] − 1 **[Keywords]** Algebraic or logical expressions **KX_RasCalculator (([R1] > 4000), [R2] − 2%([R1] > 3000), [R2] − 1%[R2])**
Example 2	**[Description] Raster** computation **[Keywords]** Algebraic or logical expressions **KX_RasCalculator(([R1] + [R2] + [R3] + [R4])/4)**
Number + function	*C002 Judgment matrix integration*
Keyword	**KX_Matrix**
Function description	The two raster layers are integrated according to the judgment matrix
Input	Raster layer (row), raster layer (column)
Output	Raster layer
Control parameters	**[Brief description]** [Polylineine description]	[Column description]	[Polyline 1 description]	[Polyline 2 description] |… The description information is expressed by numerical value, and each value is separated by ","

(continued)

Table 4.7 (continued)

Example	[Description] Integrate judgment matrix [Keywords] Raster layer (row), raster layer (column) ([5,4,3,2,1]|[5,4,3,2,1]|[3,3,3,2,2]|[3,2,2,2,1]|[2,2,2,2,1]|[2,2,2,1,1]|[1,1,1,1,1])
Number + function	*C003 Multi-layer logic integration*
Keyword	**KX_Matrix2**
Function description	Logical integration of multiple raster layers
Input	Raster layer list
Output	Raster layer
Control parameters	[Brief description] Default value | Condition description 1 | Condition description 2 | Condition description 3 |... [Description] Default value: the default value is used when the listed conditions are not met Condition description format: layer 1 value, layer 2 value, layer n value, result value. Between layers is "and" operation. When there are many control parameters, the text file can be used. The format of the text file is as follows: Line 1: default Line 2: layer 1 value, layer 2 value, layer N value, result value (between layers is and operation) Line n: ...
Example	[Description] Multi-layer logic integration [Keywords] Default value | Condition description 1 | Condition description 2 | Condition description 3 |... **KX_Matrix2(MAXTRIX2.txt)**
Number + function	*C004 Take the maximum value*
Keyword	**KX_Max**
Function description	Take the maximum value of all raster layers
Input	Raster layer list
Output	Raster layer
Control parameters	None
Example	[Description] Take the maximum value [Keywords] **KX_Max**
Number + function	*C005 Take the minimum value*
Keyword	**KX_Min**

(continued)

Table 4.7 (continued)

Function description	Take the minimum value of all raster layers
Input	Raster layer list
Output	Raster layer of end results
Control parameters	None
Example	**[Description]** Take the minimum value **[Keywords]** **KX_Min**
Number + function	*C006 Raster deduction*
Keyword	**KX_RasErase**
Function description	If the pixels in the subtracting layer are non-empty or non zero, the pixels in the corresponding position in the subtracted layer will be set to zero
Input	Subtracted raster layer, subtract raster layer
Output	Raster layer
Control parameters	None
Example	**[Description]** Raster deduction **[Keywords]** **KX_RasErase**
Number + function	*C007 Raster intersection*
Keyword	**KX_RasIntersection**
Function description	If the value of a pixel in all raster layers is not null or zero, the pixel is set to one
Input	Raster layer list
Output	Raster layer
Control parameters	None
Example	**[Description]** Raster intersection **[Keywords]** **KX_RasIntersection**

4.4.4 Keywords for Big Data Acquisition Class

Big data acquisition keywords mainly refer to keywords that use the Internet to obtain interest points, interest lines, interest planes and other operations, and their usage methods are shown in Table 4.8.

Table 4.8 Keywords for big data acquisition

Number + function	*D001 Get POI*
Keyword	**KX_BD_GetPOI**
Function description	Get POI data in input vector layer range from AutoNavi map
Input	Vector layer
Output	Vector point layer
Control parameters	**[Brief description]** Key| POI type list **[Description]** Key: the mapkey can be applied on the AutoNavi open website platform (https://lbs.amap.com) POI type list: can be downloaded from the following website, https://lbs.amap.com/api/webservice/download
Example	**[Description]** Download POI **[Keywords]** Key| POI type list **KX_BD_GetPOI(fc832fe4b573b92b743401d105de91**|06,07)**
Number + function	*D002 Get line of interest*
Keyword	**KX_BD_GetLOI**
Function description	Download LOI data according to input vector layer range on AutoNavi map website
Input	Vector layer
Output	Vector polyline layer
Control parameters	**[Brief description]** Key| LOI type list **[Description]** Key: see **KX_BD_GetPOI**
Example	**[Description]** Download LOI **[Keywords]** Key| LOI type list **KX_BD_GetLOI(fc832fe4b573b92b743401d105de91**|01,02)**
Number + function	*D003 Get area of interest (AOI)*
Keyword	**KX_BD_GetAOI**
Function description	Download AOI data of specified code on AutoNavi map website
Input	Vector layer
Output	Vector polygon layer
Control parameters	**[Brief description]** Key| Region code list **[Description]** Key: see **KX_BD_GetPOI** Region code list: Get it on the AutoNavi map website
Example	**[Description]** Download AOI **[Keywords]** Key| City code list **KX_BD_GETAOI (fc832fe4b573b92b743401d105de91**|0537)**

4.4.5 Keywords for Output Class

Output keywords are mainly common keywords related to result output (including statistical charts and reports), and their usage methods are shown in Table 4.9.

Table 4.9 Table of output class keywords

Number + function	*E001 Clip*
Keyword	**KX_Clip**
Function description	Clipping raster or vector data with vector layer
Input	Clipped layer, clip layer
Output	Raster or vector layer
Control parameters	None
Example	**[Description]** Clip **[Keywords]** **KX_Clip**
Number + function	*E002 Make thematic map*
Keyword	**KX_Mapping**
Function description	Make thematic map, support multi-layer replacement
Input	Raster or vector layer list
Output	**Thematic map**
Control parameters	**[Brief description]** Replace list I Background list I Resolution I {Template} I {Drawing range} **[Description]** Replace List: the layer style representation list in the thematic map template file (see Appendix). The input layer needs to replace the data source in the corresponding style Background List: only layers that are not replaced are displayed Resolution: the resolution of the map output Templates: include 1 and 2, which refer to thematic and questionable point map template respectively. By default, it refers to a thematic map template Drawing range: Map display range is expressed in three ways: "EmptyI* IPositioning Sentence", which respectively represents "no need to change the display rangeI zoom to the range of the first layer I zoom to the specified element"
Example 1	**[Description]** Make thematic map **[Keywords]** Replace list I Background list I Resolution I {Template} I {Drawing range} **KX_Mapping (Three area space IBoundaryI200)**

(continued)

Table 4.9   (continued)

Example 2	**[Description]** Make thematic map (questionable point map) **[Keywords]** Replace list I Background list I Resolution I {Template} I {Drawing range} **KX_Mapping (MARK, S3I I200I2I*)**
Number + function	*E003 Make thematic atlas*
Keyword	**KX_Atlas**
Function description	Produce thematic atlas and output it in PDF format
Input	Map list(Cover.pdf, Map1.pdf, Map2.pdf, ...)
Output	Atlas (Pdf format)
Control parameters	None
Example	**[Description]** Make thematic atlas **[Keywords]** Map list(Cover.pdf, Map1.pdf, Map2.pdf, ...) **KX_Atlas**
Number + function	*E004 Insert thematic map*
Keyword	**KX_InsertPic**
Function description	Insert one or more thematic maps in the specified table
Input	Map list
Output	Output table No
Input control	**[Brief description]** Picture heightI {Delete content} **[Description]** Picture height: the height of the picture specified when inserting the picture, in cm Delete content: when inserting the questionable point map, the specified text should be removed from the title as the annotation information. By default, you do not need to annotate the inserted map
Example 1	**[Description]** Insert thematic map **[Keywords]** Picture heightI {Delete content} **KX_InsertPic (10)**
Example 2	**[Description]** Insert thematic map **[Keywords]** Picture heightI {Delete content} **KX_InsertPic(4Imap)**
Number + function	*E005 Statistical table*
Keyword	**KX_Statistic**
Function description	Used for statistical analysis of a single layer or multiple layers
Input	Statistical layer, statistical layer (single or multiple)
Output	Output statistics table sequence No. or txt file name

(continued)

Table 4.9 (continued)

Output parameters	[Brief description] Display field# {Unit parameter} |Counted type list | {Data filtering mode} [Description] Unit parameter: area unit (original unit is square meter) when controlling output, 1 is square meter, 10,000 is hectare, 1,000,000 is square kilometer, default is square kilometer Data filtering mode: 0 (Name), 1 (S_{11}), 2 (R_{11}), 2n − 1 (S_{1n}), 2n (R_{1n}), S-area, R-ratio
Example 1	[Description] Statistical table [Keywords] Display field# {Unit parameter} |Counted type list| {Data filtering mode} **KX_Statistic (Name1,2,3,4,5,6,7)**
Example 2	[Description] Statistical table Input Layer: [Statistical layer], [Agricultural carrying capacity], [Provincial agricultural carrying capacity] [Keywords] Display field# {Unit parameter} |Counted type list {Data filtering mode} **KX_Statistic (Name|3,2,1|0,1,2,7,8,3,4,9,10,5,6,11,12)**
Example 2 Description	2 * 3 items are counted for two layers [Agricultural carrying capacity] and [Provincial agricultural carrying capacity], and the display order is as follows: Name, V_{11}, R_{11}, V_{21}, R_{21}, …, V_{23}, R_{23}
Example 3	[Description] Statistical table Input layer: [Statistical layer], [Ecological maximum], [Ecological minimum], [Agricultural maximum], [Agricultural minimum], [Urban maximum], [Urban minimum] [Keywords] Display field# {Unit parameter} |Counted type list| {Data filtering mode} **KX_Statistic (Name|1)**
Example 3 description	The area of six statistical input layers, [Ecological maximum value], [Ecological minimum value], [Agricultural maximum value], [Agricultural minimum value], [Urban maximum value] and [Urban minimum value], which belong to counted layers type 1
Example 4	[Description] Statistical table [Keywords] Display field# {Unit parameter} |Counted type list| {data filtering mode} **KX_Statistic(name#1|1,2,3)**
Example 4 description	Counts the areas of types 1, 2 and 3 in the counted layer and outputs them in meters

4.5 Features and Applications of G Language

4.5.1 Features of G Language

Although G language is a programming language, to make it easy for more people who do not have programming thoughts to understand and master, G language has its own uniqueness compared with the existing mainstream programming languages in terms of expression form and grammar rule design, mainly reflected in the following aspects:

(1) Because G language is a domain-specific language oriented to geographical modeling, compared with traditional programming languages, it has a higher level of abstraction and pays attention to the problem of doing what rather than how to do it. For this reason, only keywords are used in G language to express the what that needs to be done in GC, and the programming contents that users are not easy to master, such as loop and judgment in traditional programming language, are eliminated.

(2) G language, a highly abstract language, shields the differences of different GIS platforms and is closer to natural language. Professionals only need to master a limited number of G language keywords used to build geographic analysis models to build their own geographic analysis models without mastering the operation of GIS platform. In this way, the application of GIS has stepped from platform-level application to language-level application, which greatly reduces the application threshold of GIS and plays an important role in the popularization of GIS.

(3) From the perspective of expression form, G language is a tabular programming language, which expresses the logic of the whole keywords are used in G language to express the what that needs to be done in the GC task through tables, replacing various restrictions on code format of traditional programming languages (such as indentation and case difference of Python language). This design not only effectively reduces the probability of errors when users write intelligent documents, but also simplifies the difficulty of document parsing by the background G language interpreter.

(4) In traditional programming languages, the parameter part of the function mixes the input, output and control parameters together for processing, which brings many difficulties to the use of the function. In G language, the input layer, output results and control parameters are placed in different columns of the table. This processing method standardizes the user's input and enables the user to easily master the use of keywords.

(5) Traditional programming languages usually use dialog boxes to set running parameters, but in G language, strings are used to set control parameters. By adopting this design, on the one hand, the repeated input of users in the GC process can be saved to achieve the purpose of fully automatic data processing, and on the other hand, the control parameters during computation can be completely retained, thus achieving the goal of "one input, multiple uses", thus greatly improving the computation efficiency of complex GC task.

(6) Because GC items and control parameters of the whole GC process are completely recorded in the [GC process table], When the quality of the GC results is checked, the problems of the computation items and parameters in the GC process can be found by comparing the original computation items and parameters, thus it can be clearly understood that the GC results are obtained

through the geographical analysis process of what. This can not only enable business staff to have a clear idea of the processing results, but also effectively prevent the occurrence of falsification of GC results, which is a difficult problem to solve for the existing GC pattern.

(7) In traditional programming languages, code annotations have always been a problem for programmers, although the code comments do not affect the operation of the program, however, standard code annotation is of great significance to the maintenance of programs. However, because adding annotations will take up a lot of time for programmers, many programmers have resistance to code annotation. In addition, there is no rigid stipulation on the position of code annotation in traditional programming languages; code annotation has always been a long-standing problem in the programming field. However, in G language, it is a programming environment based on MS word or Kingsoft WPS documents. The process of Building GC model with G language is the process of writing MS word or Kingsoft WPS document, which is a standardized description of the construction principle, method and process of GC model. For each operation step, there is space for annotation to ensure people's understanding and knowledge of computational logic.

4.5.2 Applications of G Language

Because G language has the characteristics of natural language, it can be mastered by more people and greatly reduces the application threshold of GIS. Therefore, it has a wide application prospect. The application of G language mainly includes the following three aspects.

1. Construct a business platform

Because G language is easy to learn and implement, ordinary business staff can complete daily data processing and data analysis according to business needs, such as Dual-evaluation business of territorial and spatial planning, environmental evaluation business or spatial data processing business in MS word or Kingsoft WPS environment. This can not only ensure the quality of data analysis or data processing results and improve work efficiency, but also make business work more standardized and orderly.

2. Construct research and exploration platform

With the increasing maturity and popularization of GIS technology, more and more researches are related to spatial analysis. However, it is regrettable that most technicians use manual GC pattern to process and analyze data. As previously analyzed, this manual GC pattern is inefficient and cannot guarantee the quality of processing results. The research and exploration platform constructed by G language can effectively integrate the research theory, GC process and GC results in MS word or Kingsoft WPS environment, and become a truly verifiable, traceable and reusable knowledge system. Researchers can easily verify and compare different theories and methods on this research platform, and test the influence of different parameter settings on the results, thus providing many quantitative analyses data support for the research.

3.   **Construct GIS teaching experiment platform**

At present, many colleges and universities have set up GIS or remote sensing experimental courses, hoping to enable students to master the operation of GIS or remote sensing software by completing a series of experiments (Tang and Yang 2007; Deng 2010). However, due to the complexity of GIS or remote sensing software, students usually spend a lot of time following the steps introduced in the tutorial to complete the experiment. In this process, students all focus on how to find tools and how to set parameters, but do not have more energy to think about the principle or why. This way of learning GIS is debatable. What's more, there are many kinds of mainstream GIS or remote sensing software. How do students choose?

For G language provides a spatial analysis method unrelated to the operation of GIS software, students can use G language to build their own GC model in MS word or Kingsoft WPS environment to complete the internship task. In this platform, students will no longer stick to the operation of step-by-step dialog box, but will focus on how to use spatial analysis to solve problems, which will effectively cultivate students' geographical thinking ability. At the same time, the experimental report completed by the students is a verifiable and reusable MS word or Kingsoft WPS documents, whether it is for teachers to check students' homework, or for students to use their own knowledge in the future is very useful.

References

Deng S (2010) ENVI remote sensing image processing methods. Science Press, Beijing (in Chinese)
Tang G, Yang X (2007) ArcGIS geographic information system spatial analysis experimental course. Science Press (in Chinese)

Part II
The Practice of the New GC Pattern

Chapter 5
Implementation of G Language

5.1 Implementation Strategy

If any programming language is to be applied, it is still far from enough to have a good syntax design. It is also necessary to develop a corresponding compiler or interpreter, and at the same time, it is also necessary to have the support of a code editing, debugging, and running environment.

5.1.1 Program Operation Mode

1. Modalities for the execution of the procedure

At present, advanced programming languages are divided into compilation and interpretation languages, which operate in compilation mode and interpretation mode respectively.

Compile mode is to translate advanced language programs into low-level language programs through translation programs, and carry out data processing with the support of runtime library programs (as shown in Fig. 5.1). C, C++, FORTRAN, PASCAL, C # and VB .net are all compiled programs.

The interpretation mode is to run the advanced language program while interpreting through the interpreter (as shown in Fig. 5.2) without generating executable files. LISP, ML, Prolog, Smalltalk, and currently popular Python are all explanatory languages.

The comparisons between the two languages are shown in Table 5.1.

2. Execution mode of G language

As a domain-specific language for end users, G language is positioned as an explanatory language and runs in an explanatory execution mode. There are mainly the following two considerations:

© Surveying and Mapping Press 2021
W. Zhou, *A New GeoComputation Pattern and Its Application in Dual-Evaluation*,
https://doi.org/10.1007/978-981-33-6432-5_5

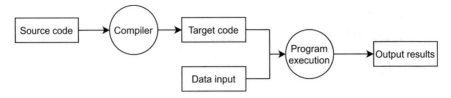

Fig. 5.1 Compile execution mode of program

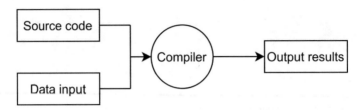

Fig. 5.2 Explanation and execution of the program

Table 5.1 Comparisons of explanatory and compiled languages

Contrast item	Compiled language	Explanatory language
Features	When compiling, it is directly compiled into what the machine can execute (.exe .dll .ocx), and compilation and execution are separated	Explanatory language programs do not need to be compiled, and explanatory language is translated only when the program is running
Advantages	1. Since translation is only done once and no translation is required when running, the program execution efficiency of compiled language is high 2. The compiled program cannot be modified and has good confidentiality	1. It has good portability. If there is an interpretation environment, it can run on different operating systems 2. With good flexibility, code can be run after modification, no compilation process is required
Disadvantages	1. The code needs to be compiled before it can run 2. The portability is poor and can only run on compatible operating systems	1. Running needs to explain the environment, which is slower than compiling and takes up more resources 2. It is inefficient to translate once every time it is executed 3. The code efficiency is low, because not only the space should be allocated to the user program, but also the interpreter itself takes up valuable system resources

Fig. 5.3 Execution schematic of G language interpretation

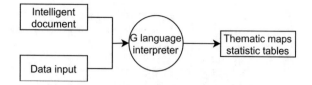

(1) The original intention of creating G language is to provide users with a flexible and convenient method for building GC models, and the models can be adjusted and optimized at any time according to the needs of actual applications. Based on this consideration, G language should adopt interpretation type instead of compilation type.

(2) G language is a programming language oriented to the GC context. Its GC capability is based on the existing GIS platform. Using interpretation mode can flexibly, effectively, and conveniently integrate these resources.

As an explanatory language, the execution mode of G language is to interpret the intelligent documents written by G language (source code from the perspective of traditional programming) through G language interpreter, and call the functions of other GIS platforms to carry out specific data processing to form various thematic maps and statistical tables. Figure 5.3 shows the schematic diagram of G language interpretation and execution.

5.1.2 Integrated Development Environment

1. Analysis of existing integrated development environment

Different software is used to process each stage of earlier program design. For example, the document processing software is used to edit the source program first, then the link program is used to connect functions and modules, and then the compiler is used to compile. Developers must switch back and forth between several kinds of software. Later programming software usually integrates editing, compiling, debugging and other functions into an environment (called Integrated Development Environment, IDE), which can greatly facilitate the development of the system.

At present, common IDEs include Microsoft's Visual Studio series (see Fig. 5.4), Borland's C++ Builder, JetBrain's PyCharm (see Fig. 5.5), and Eclipse of the Eclipse Foundation.

2. G language integrate development environment

For G language, it is a unique programming language. It cannot use the existing IDE and can only be redeveloped. However, it is not easy to develop a fully functional IDE. Therefore, according to the grammatical characteristics and application scenarios of G language (used by non-programmers), the document processing software MS Word

Fig. 5.4   Visual studio 2012 integrated development environments

Fig. 5.5   PyCharm 5 integrated development environments

or Kingsoft WPS is selected as IDE developed as G language, based on the following four considerations.

(1)   MS Word is currently the most popular document editing software in the world (as shown in Fig. 5.6). Almost every computer equipped with Windows operating system is equipped with MS Word. It can be said that MS Word is the most widely used learning and working platform in the world. Using MS Word as an intelligent document editor can greatly reduce the burden of non-programmers

Fig. 5.6 MS Word document processing software

learning and using G language. WPS is a document processing software independently developed by Kingsoft Company of China (as shown in Fig. 5.7), which is compatible with MS Word, but the memory occupation of WPS is relatively small, and the use of personal version is free. At present, many government agencies and companies in China use the software.

(2) G language takes tables as the basis of grammatical rule expression, no matter the expression of basic parameters, the registration of GC tasks, the description of GC process and the expression of GC results are all realized through tables,

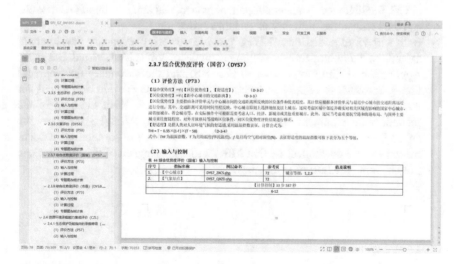

Fig. 5.7 Kingsoft WPS document processing software

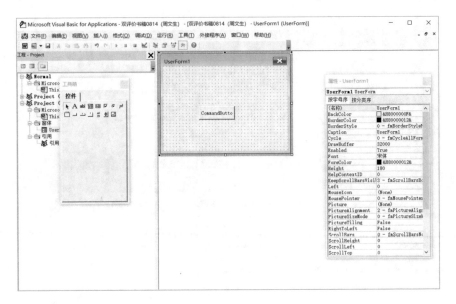

Fig. 5.8 MS Word VBA editor interface

while MS Word and Kingsoft WPS software have strong table processing capability, which can provide great convenience for the compilation of intelligent documents.

(3) Using MS Word or Kingsoft WPS software as the integrated development environment of G language can effectively integrate the model description, GC process and GC results, avoid switching between several systems In the process of GC, facilitate the management, maintenance, and application of GC models, and thus reduce the difficulty of system application.

(4) Both MS Word and Kingsoft WPS provide powerful secondary development capability, including the development mode of embedded Visual Basic for Applications (VBA) (as shown in Fig. 5.8) and the secondary development library for other development languages. All of these provide rich means for manipulating WORD documents or WPS documents. With the help of the secondary development ability of MS Word or Kingsoft WPS software, the debugging and running functions of intelligent documents (or intelligent documents) can be developed, so that MS Word and Kingsoft WPS software can truly become the IDE of G language.

5.2 G Language Interpreter

5.2.1 Overall Structure

In addition to its own grammar rules for compiling intelligent documents, G language also needs a matching G language interpreter (or back-end service system) for interpreting and executing the contents of intelligent documents.

The whole architecture of G language interpreter is shown in Fig. 5.9.

G language interpreter is mainly composed of grammar analysis module and keyword execution module, the syntax analysis module parses intelligent documents with the support of document secondary development library and G language syntax rules, while the keyword execution module completes GC operation with the support of GIS secondary development library and keyword execution library according to the analysis results of the syntax analysis module. Among them, the document secondary development library refers to the development library of MS Word or Kingsoft WPS, and the GIS secondary development library refers to the secondary development library of ArcGIS, SuperMap, MapGIS and open-source GIS. The keyword execution library implements the keyword operation defined by G language.

It should be noted that since the keyword execution library is based on the existing GIS platform, from the perspective of users, the G language interpreter realizes the goal of crossing GIS platforms, that is, the end users only need to know the usage method of keywords in G language, and do not need to know the specific details of how keywords are executed in different GIS platforms. This means that although the G language interpreter is implemented based on the existing GIS platforms, the end users do not need to be familiar with the operation of these GIS platforms, thus reducing the burden for users to learn GIS, lowering the application threshold of GIS, enabling more end users to easily use the complex functions of GIS through G language, and further promoting the popularization of GIS.

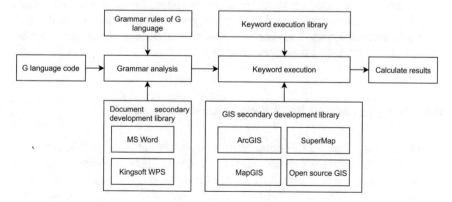

Fig. 5.9 Architecture system of G language interpreter

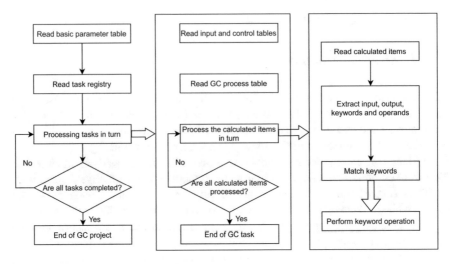

Fig. 5.10 Processing logic for G language interpreter

5.2.2 Operation Logic

The specific execution logic of analyzing intelligent documents by using G language interpreter is shown in Fig. 5.10.

G language first reads the [Basic parameter table] and [GC task registry] in the intelligent document respectively, and then processes each GC task that needs to be processed.

When processing a GC task, the input layer information and computation control information in the [Input and control table] are first read, and then the GC item information is sequentially extracted from the [GC process table] for processing according to the computation control information. When processing each GC item, it is necessary to analyze keywords, input items, output items, control parameters and other information according to G language grammar rules, and then match with keyword execution library. The matched keywords complete GC operation under the support of GIS secondary development library.

5.3 G Language Debug Operation Subsystem

Although G language directly uses document processing software MS Word or Kingsoft WPS as the integrated development environment, these two sorts of software only provide document editing function, and the debugging and operation functions of G language need to be realized by means of the secondary development capability provided by MS Word or Kingsoft WPS.

5.3.1   Functional Model Design

Referring to the system debugging and running functions of the general integrated development environment (such as JetBrain's PyCharm) and considering the specific needs of intelligent document debugging and running, a G language debugging and running subsystem shown in Fig. 5.11 is designed, which is called G language IDE.

The whole subsystem includes five functional modules: Overall control, Basic evaluation, Extended evaluation, Map analysis and About.

(1) Overall control: Used to control the overall functions of the system, such as setting the operating parameters of the system, checking the legality of intelligent documents, exporting workspace, and fully automatic computation of GC tasks, etc., which correspond to the functions of environment setting, document maintenance and automatic computation respectively.

(2) Basic evaluation: For more complex GC projects, there are usually three parts: individual evaluation, integrated evaluation, and comprehensive evaluation. Here, individual evaluation, integrated evaluation and comprehensive evaluation are set up to correspond to them. It should be noted that only the functional framework is set here, and the specific content needs to be configured according to the situation of GC tasks.

(3) Extended evaluation: This module is a supplement to the basic evaluation functional module, and non-basic evaluation GC tasks can be configured in this functional module. At present, it includes optional evaluation and comparative analysis. Like the basic evaluation module, the specific content of this module also needs to be configured according to the content of GC tasks.

(4) Map analysis: Spatial data visualization is an important content in GC. This module directly calls the background GIS platform to complete spatial data visualization and editing of thematic map templates, which correspond to the two functions of visual analysis and mapping template respectively. In addition, the output of batch thematic maps is also an important part of GC. Therefore, a map analysis function is set up, and the specific content of this function also needs to be configured according to the content of GC tasks.

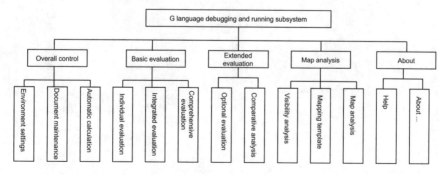

Fig. 5.11 Function model of G language debugging subsystem

(5) About: Provide help information and copyright information on the use process of the system, mainly including help and about two functions.

5.3.2   User Interface Customization

The interface of G language debugging and running subsystem is mainly customized through the custom functional area technology provided by MS Word or Kingsoft WPS. Custom Ribbon is a user interface design technology initiated by Microsoft Office 2007 to help users quickly define their own Ribbon. In the definition ribbon, commands commonly used by users are organized in several logical groups, which are organized under tabs.

The following is an example of MS Word 2013 to introduce the customization method of G language debugging and running subsystem interface.

(1) Click the MS Word "File" tab and click "Options" to pop up the "Word Options" dialog box (as shown in Fig. 5.12);
(2) Click "Custom Functional Area", select "Main Tab" in the "Custom Functional Area" drop-down box on the right, and modify the name of the custom functional area to [G language IDE] in the "Custom Functional Area" list box;

Fig. 5.12 Customize the interface in MS Word 2013

Fig. 5.13   G language debug run subsystem interface

(3) Add 5 tabs such as [Overall control], [Basic evaluation], [Extended evalua-tion], [Map analysis] and [About] in turn under the [G language IDE] custom functional area.

(4) Add groups under each newly added tab, such as [Environment setting], [Docu-ment maintenance] and [Automatic computation] groups in the [Overall control] tab.

(5) Select "Macro" in the drop-down box on the left of the "Word Options" dialog box. At this time, user-defined function functions will be listed in the corre-sponding list box. Users can "add" or drag the corresponding function to the corresponding group, such as adding the [Environment setting] function in the list box in the [Environment setting] group.

After the above settings are completed, the [G Language IDE] functional area shown in Fig. 5.13 will be formed.

5.3.3   Operation Logic Implementation

The implementation logic associated with the interface of G language debugging and operation subsystem is implemented through VBA provided by MS Word or Kingsoft WPS. VBA is a macro language of Microsoft Visual Basic. VBA can be used to conveniently associate GC tasks with G language interpreter, thus realizing the operation and debugging of intelligent documents (as shown in Fig. 5.14).

5.4   Compiling of Intelligent Documents

5.4.1   Intelligent Documentation

Intelligent document is a kind of hybrid document written in MS Word or King-soft WPS document by using grammar rules of G language. The reason why it is called intelligent document is that, unlike the general MS Word or Kingsoft WPS documents, the GC process described in the document can be recognized by the G language interpreter and the described operations can be performed in the back-ground. From the perspective of general programming, intelligent document can also

Fig. 5.14 Operation and debugging interface of GC task

be regarded as the program code of G language. However, considering that most users of G language are non-programmers, coding is called the compiling of intelligent documents.

Compared with the coding of other programming languages, the compiling of intelligent documents has specific and distinctive features, which are mainly reflected in the following three aspects:

(1) The compiling of intelligent documents is carried out in familiar MS Word or Kingsoft document, so that the powerful editing function of MS Word or Kingsoft WPS can be fully utilized to organize the compiling of intelligent documents reasonably and effectively. The overall structure of intelligent documents can also be clearly understood through the navigation window and quickly navigated to the required location (as shown in Fig. 5.15).

(2) The grammar rules of G language are mainly form filling rules, i.e. The compiling of intelligent documents is mainly to fill in [Basic parameter table], [GC task registry], [Input and control table] and [GC process table] (as shown in Fig. 5.16). Therefore, the compiling of intelligent documents is much simpler than general programming, and end users can master it through simple learning.

(3) Intelligent documents are mainly described by business logic, described by natural language, and limited G language syntax rules, and mainly describes the problem of doing what. The specific details of how to do it are hidden in the background G language interpreter through encapsulation technology. Therefore, there are no complicated structures such as loops and judgments in general programming languages in intelligent documents. In this way, the end user only needs to be familiar with the business logic of a GC task (such as land

Fig. 5.15 Navigation for MS Word

resource evaluation) to realize the smooth conversion of business logic from natural language description to implementation logic.

5.4.2   Organization of Intelligent Documents

Compared with general program codes, the structure of intelligent documents is relatively simple. It is mainly composed of three parts, namely, basic parameters, GC task registration and GC tasks. The basic parameters are equivalent to global variables in general programs, the GC task register a function declaration, GC tasks the definition of function bodies, which is the core content of intelligent documents. When there are many GC tasks, relevant GC tasks can be organized and managed through GC task groups (equivalent to modules in general programming languages) according to the general programming ideas. Figure 5.17 shows the general framework of intelligent documents.

The structure of the above intelligent documents can be realized by using the title style function in MS Word or Kingsoft WPS. The specific operations are as follows:

(1) Set the first-level titles of [Basic parameters], [GC task registration] and [GC task] in the MS Word document;

(2) Add a [Basic parameter table] under the first-level title of [Basic parameters] according to G language rules, insert a [GC task registry] under the first-level title of [GC task registration], and respectively set several second-level titles of GC task groups under the first-level title of [GC task];

(3) Set a plurality of GC task three-level headings under that two-level heading of each GC task group;

7.	Correct [Importance of ecosystem services 3] with @	[Key biodiversity conservation areas R] [Importance of ecosystem services 2]	**[Description]** Raster calculator If[Key biodiversity conservation areas R] = 1, then [Importance of ecosystem services] = 3 **[Keywords]** **KX_RasCalculator (([R1] ==1), 3% [R2])**	[Importance of ecosystem services] DX1_ STZYX	
8.	Rasterization @	[Sensitivity of soil erosion]	**[Description]** Rasterization **[Keywords]** Operation field | Grid size **KX_FeatureToRaster (DJ)**	[Sensitivity of soil erosion R] DX1_STLS	
9.	Rasterization @	[Desertification sensitivity]	**[Description]** Rasterization **[Keywords]** Operation field | Grid size **KX_FeatureToRaster (DJ)**	[Desertification sensitivity R] DX1_SMH	
10.	Rasterization @	[Sensitivity of rocky desertification]	**[Description]** Rasterization **[Keywords]** Operation field | Grid size **KX_FeatureToRaster (DJ)**	[Sensitivity of rocky desertification R] DX1_SHMH	
11.	Rasterization @	[Coastal erosion sensitivity]	**[Description]** Rasterization **[Keywords]** Operation field | Grid size **KX_FeatureToRaster (DJ)**	[Coastal erosion sensitivity R] DX1_HAQS	
12.	Calculate #	[Sensitivity of soil erosion R] [Desertification sensitivity R] [Sensitivity of rocky desertification R] [Coastal erosion sensitivity R]	**[Description]** Take the maximum value **[Keywords]** **KX_Max**	[Ecological sensitivity] DX1_ STMGX	
13.	Make thematic map #	[Importance of ecosystem services]	**[Description]** Make thematic map **[Keywords]** Replace list | Background list | Resolution | template| Positioning statement **KX_Mapping(S1|Boundary|200)**	[Thematic map of importance of ecosystem services] DX1_STZYX.emf	

Fig. 5.16 Exemplary completion of GC process table

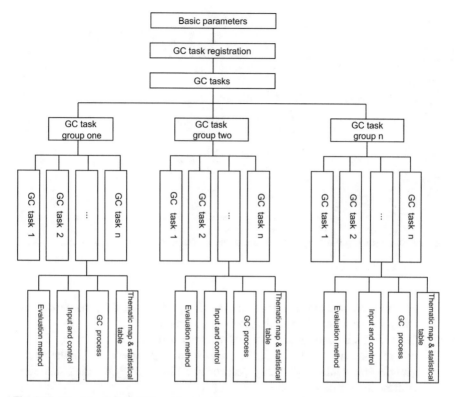

Fig. 5.17 Framework for intelligent documents

(4) Under the three-level titles of each GC task, four levels of titles are set up, including [Evaluation method], [Input and control], [GC process] and [Thematic map & statistical table]. And insert [Input and control Table], [GC process table], [Thematic map & statistical table] under [Input and control], [GC process] and [Thematic map & statistical table] respectively.

Figure 5.18 shows the effect of the intelligent document framework in Word documents.

5.4.3 GC Task Edit

1. Increase the number of GC tasks

Adding a GC task requires the following three steps:

(1) A third-level title GC task unit (such as land resource evaluation) is added to the second-level title (such as individual evaluation group) of the GC task group. The first GC task unit includes four parts: evaluation method, input and control,

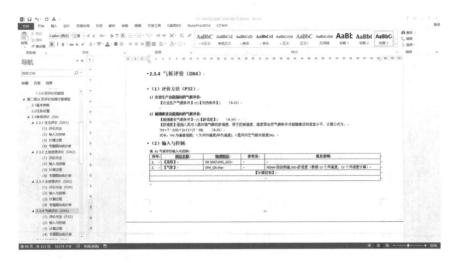

Fig. 5.18 Intelligent document framework

GC process and thematic map-statistical chart. After that, the contents of these four parts are filled in according to G language rules.

(2)   The GC task is registered in the [GC task registry].

(3)   Edit the menu configuration file Menu.txt in the system directory or the intelligent document sibling directory (as shown in Fig. 5.19) to add the registration information for the GC task.

2.   Delete the GC task

Corresponding to the addition of GC tasks, deleting a GC task also requires 3 steps.

(1)   Delete the GC task unit from the MS Word or Kingsoft WPS document;

(2)   Delete the relevant information of the GC task from the [GC task registry];

(3)   Delete the GC task registration information from the menu configuration file Menu.txt.

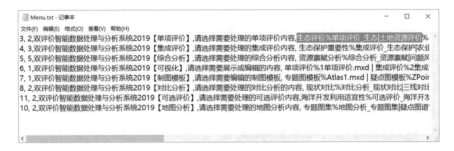

Fig. 5.19 Adding GC tasks to the menu file

3. **Modify GC tasks**

If you modify the name of the GC task, you need to make corresponding adjustments in the [GC task registry] and the menu configuration file. If you only modify the information content in the [GC process table], you do not need to adjust the [GC task registry] and the menu configuration file.

5.4.4 Notes

A note is the explanation of the program codes. Its purpose is to make others and themselves understand the meaning expressed by the code. Different programming languages usually adopt different annotation methods. For example, Python language provides single-line annotation and multi-line annotation, where single-line annotation starts with #, for example:

```
# This is a note.
print("Hello, G language!")
```

There is no specific annotation identifier in G language, but as an end-user-oriented programming language, G language provides a variety of annotation description methods to describe GC tasks to the greatest extent.

1. **Notes outside the table**

The table in the intelligent document is an important element to realize the interaction with the background G language interpreter, and the description content outside the table has no effect on the G language interpreter. Therefore, the GC task can be fully explained outside the table by means of pictures and texts, as shown in Fig. 5.20, which is an example of the explanation of the water resources evaluation model.

2. **Notes in the table**

The tables in the intelligent document include [Basic parameter table], [GC task registration table], [Input and control table] and [GC process table]. These four types of tables all have description columns to explain various data or operations. Similarly, the G language interpreter does not do anything about the contents in the description column. The [Description] column in the [GC process table] shown in Fig. 5.21 is a description of each GC item.

In addition, the information in the non-description column of the table can also be modified by making full use of the text modification function in MS Word or Kingsoft WPS document. For example, in the [GC process table], the red font can be used in the [Input] column to represent the input layer set by the GC task, and the blue font can be used in the [Output] column to represent the final output layer of the GC task.

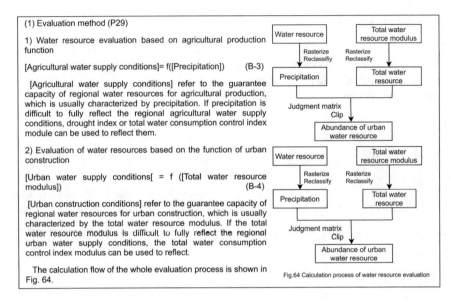

(1) Evaluation method (P29)

1) Water resource evaluation based on agricultural production function

[Agricultural water supply conditions]= f([Precipitation])   (B-3)

[Agricultural water supply conditions] refer to the guarantee capacity of regional water resources for agricultural production, which is usually characterized by precipitation. If precipitation is difficult to fully reflect the regional agricultural water supply conditions, drought index or total water consumption control index module can be used to reflect them.

2) Evaluation of water resources based on the function of urban construction

[Urban water supply conditions[= f ([Total water resource modulus])   (B-4)

[Urban construction conditions] refer to the guarantee capacity of regional water resources for urban construction, which is usually characterized by the total water resource modulus. If the total water resource modulus is difficult to fully reflect the regional urban water supply conditions, the total water consumption control index modulus can be used to reflect.

The calculation flow of the whole evaluation process is shown in Fig. 64.

Fig.64 Calculation process of water resource evaluation

Fig. 5.20 Extra-table note description

5.5 Debugging and Running of Intelligent Document

5.5.1 Debugging Method

The intelligent document can be run or debugged in the following four modes.

1. **Single task operation mode**

In this mode, the system will perform the corresponding operations row by row according to the contents in the GC process table. The operation of this mode only needs to select the GC task to be executed in the menu bar and set the content of the next row of [GC control] to blank in the [Input and control table] of the corresponding GC task (as shown in Fig. 5.22).

2. **Single step operation mode**

When debugging or running a GC task, to run a GC item in the GC process table separately, you can set the [GC control] content in the [Input and control table] to the serial number of the row where the GC item is located.

3. **Multi-step operation mode**

When debugging or running a GC task, to run several computation steps in the [GC process table], you can set the contents of [GC control] according to the rules described in Sect. 4.2.3. If you want to run the GC items in steps 1-3, fill in 1-3 or 1:3, and if you want to run the GC items after step 4, fill in 5- or 5:, as shown in Fig. 5.23 the example is to skip the 4 GC items during operation.

Step	Operating instructions	Input	Operation	Output	Description
1.	Interpolate active accumulated temperature	[Weather station]	[Description] Interpolation [Keywords] Interpolation field | Interpolation method (IDW, KRI) | Resolution 20 KX_Interpolation(HDJW|IDW)	[Photothermal conditions 1] DX4_GRTJ1	[Active accumulated temperature]
2.	Modify with DEM	[Photothermal conditions 1] [Height]	[Description] Raster calculator [Keywords] KX_RasCalculator([R1]-[R2]/100*0.6)	[Photothermal conditions 2] DX4_GRTJ2	
3.	Reclassification of [photothermal condition 2]	[Photothermal conditions 2]	[Description] Reclassification (numerical range) [Keywords] KX_Reclass(1: <1500| 2: 1500-4000| 3:4000-5800 |4:5800-7600 |5: >=7600)	[Photothermal conditions 3] DX4_GRTJ3	
4.	Clip @ with the range layer	[Photothermal conditions 3] [Range layer]	[Description] Clip [Keywords] Clipped layer, clip layer KX_ExtractByMask	[Photothermal conditions] DX4_GRTJ	5（Excellent）/4（Good）/3（Medium）/2（Poor）/1（Bad）
5.	Generate [Comfort 1] with temperature and humidity of weather stations	[Weather station]	[Description] Comfort [Keywords] Name, comfort name, meteorological TXT file (name, 12 months temperature list, 12 months humidity list) KX_SSD (Name, SSD, SSD.txt)	[Comfort 1] DX4_SSD1.shp	KX_SSD (name, SSD, SSD.txt,1)
6.	Generate # [Comfort 2]	[Comfort 1]	[Description] Interpolation [Keywords] Interpolation field | Interpolation method (IDW, KRI) | Resolution 20 KX_Interpolation (SSD|IDW)	[Comfort 2] DX4_SSD2	
7.	Generate # [Comfort 2]	[Comfort 2]	[Description] Reclassification (By range) [Keywords] KX_Reclass (1: <35| 2: 35-40| 3:40-45 |4:45-50 |5:50-70 |6:70-75 |7:75-80 |8:80-85 | 9: >=85)	[Comfort 3] DX4_SSD3	

Fig. 5.21 Notes in the table of GC process

Serial number	Layer Name	Physical layer	Reference page	Value and description
1.	[Height]	DX2_GC		50m * 50m in country and province, 20m * 20m or 30m * 30m in city or county
2.	[Soil texture]	DX2_TRZD3		Percentage of silt content in soil
[GC control]				

Fig. 5.22 Input and control table settings at single task runtime

Serial number	Layer Name	Physical layer	Reference page	Value and description
1.	[Height]	DX2_GC		50m * 50m in country and province, 20m * 20m or 30m * 30m in city or county
2.	[Soil texture]	DX2_TRZD3		Percentage of silt content in soil
[GC control]				
1-3,5:				

Fig. 5.23 Multi-step runtime input and control table settings example

4. Automatic computation mode

After the debugging of a single GC task is completed, the automatic computation mode can be used to carry out batch computation on the GC tasks selected by the GC project. During the specific operation, the [Calculate or not] column of the task to be computed is set to "Y" in the [GC task registry], and then the required GC task can be fully completed through [G language IDE], [Overall control] and [Automatic computation] (as shown in Fig. 5.24).

5.5.2 Debugging of Intelligent Documents

For G language, the debugging of its code is the debugging of GC tasks or GC models. The main content of debugging is the [GC process table] of each GC task. The single-step mode or multi-step mode can be adopted for specific debugging. During debugging, users can add and delete GC items at the required positions, and at the same time they can modify the control parameters of keywords (as shown in Fig. 5.25).

Since the final output results in the GC task are displayed in the [Thematic map-statistical table], if visual analysis is needed for the intermediate results, the external GIS platform can be opened through [G language IDE], [Map analysis] and [Visual analysis], and the layer contents specified in the output column in the [GC process table] can be added (as shown in Fig. 5.26).

Serial No.	Task name	Calculate or not	Workspace	Table location (input and control table)
1.	Individual evaluation_ecology	Y	DX/DX1	Table 6
2.	Individual evaluation_land resource	Y	DX/DX2	Table 12
3.	Individual evaluation_water resource	Y	DX/DX3	Table 18
4.	Individual evaluation_climate	Y	DX/DX4	Table 24
5.	Individual evaluation_environment	Y	DX/DX5	Table 30
6.	Individual evaluation_disaster	Y	DX/DX6	Table 40
7.	Individual evaluation_comprehensive advantage degree (provincial level)	Y	DX/DX71	Table 46
8.	Individual evaluation_comprehensive advantage (city and county)	Y	DX/DX72	Table 50
9.	Integrated evaluation_ecological protection		JC/JC1	Table 58
10.	Integrated evaluation_agricultural production		JC/JC2	Table 62
11.	Integrated evaluation_urban construction		JC/JC3	Table 66
12.	Integrated evaluation_agricultural scale		JC/JC4	Table 72
13.	Integrated evaluation_urban scale		JC/JC5	Table 76
14.	Comprehensive analysis_resource endowment		ZH/ ZH1	Table 80
15.	Comprehensive analysis_problem risk		ZH/ZH2	Table 85
16.	Comprehensive analysis_agricultural potential		ZH/ ZH3	Table 88
17.	Comprehensive analysis_urban potential		ZH/ ZH4	Table 94
18.	Comprehensive analysis_scenario analysis		ZH/ ZH5	Table 100

Fig. 5.24 Settings in [Task setting table] for automatic computation

5.5.3 Output of Operation Results

1. Runtime window

During the execution of the intelligent document, a small runtime window (as shown in Fig. 5.27) will pop up to display the currently executed operation in real time. At the same time, the cursor will automatically move to the corresponding execution steps in the [GC process table], so that users can know the execution of GC tasks in time. It should be noted that the running small window only displays the relevant information of the GC process and has no interactive function.

Step	Operating instructions	Input	Operation	Output	Description
1.	Rasterization @	[Frequency of drought disaster]	[**Description**]Rasterization [Keywords]Operation field | Grid size KX_FeatureToRaster (GHPL)	[Risk of drought disaster 1] DX6_GHWXX1	
2.	Reclassification @	[Risk of drought disaster 1]	[**Description**]Reclassification (numerical range) [Keywords] KX_Reclass (1:<20|2:20-40|3:40-60|4:60-80|5:>=80)	[Risk of drought disaster] DX6_GHWXX	5（Higher）/4（High）/3（Medium）/2（Low）/1（Lower）
3.	Rasterization @	[Frequency of flood disaster]	[**Description**]Rasterization [Keywords]Operation field | Grid size KX_FeatureToRaster (HLPL)	[Flood hazard 1] DX6_HLWXX1	
4.	Reclassification @	[Flood hazard 1]	[**Description**]Reclassification (numerical range) [Keywords] KX_Reclass (1:<20|2:20-40|3:40-60|4:60-80|5:>=80)	[Flood hazard] DX6_HLWXX	5（Higher）/4（High）/3（Medium）/2（Low）/1（Lower）
5.	Rasterization @	[Frequency of low temperature and cold disaster]	[**Description**]Rasterization [Keywords]Operation field | Grid size KX_FeatureToRaster (DWLPL)	[Risk of low temperature and cold disaster 1] DX6_DWLWXX1	
6.	Reclassification @	[Risk of low temperature and cold disaster 1]	[**Description**]Reclassification (numerical range) [Keywords] KX_Reclass (1:<20|2:20-40|3:40-60|4:60-80|5:>=80)	[Risk of low temperature and cold disaster] DX6_DWLWXX	5（Higher）/4（High）/3（Medium）/2（Low）/1（Lower）

Fig. 5.25 Modification of [Control parameters] during debugging

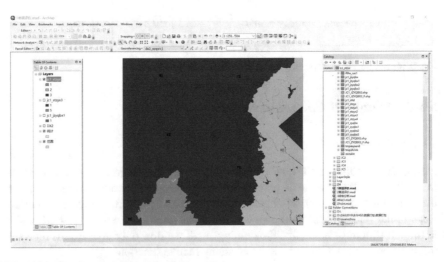

Fig. 5.26 Visual analysis of intermediate results

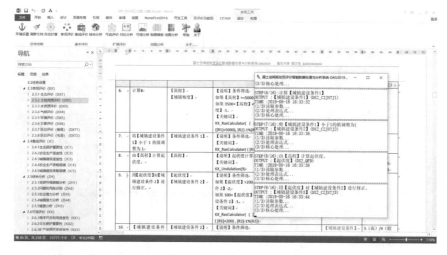

Fig. 5.27 System runtime window

2. **Runtime log**

For each operation of the system, the operation contents of each step of the operation process and the error information at runtime will be completely recorded in the log file (as shown in Fig. 5.28) to provide support for users to debug the system.

3. **Visualization of result**

For a GC task, its GC results (intermediate results and end results) have four forms, namely raster layer, vector layer, thematic map, and statistical table. These data are stored at the positions specified in the output column of [GC process table]. For users to see the results (thematic maps and statistical tables) in time, thematic maps and statistical tables can be output to the specified palaces in MS Word or Kingsoft WPS document through keywords KX_InsertPic and KX_Statistical in the GC task (as shown in Fig. 5.29).

When users need to view the intermediate GC results, they can directly call the background GIS platform and load the layer data to be analyzed through the [Visual analysis] function in the [Map analysis] tab (as shown in Fig. 5.30).

Fig. 5.28 System log file

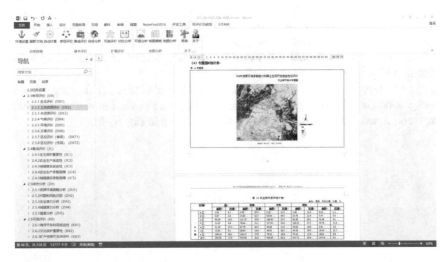

Fig. 5.29 Output of GC results

Fig. 5.30 Visual analyses of intermediate results under GIS platform

Chapter 6
Dual-Evaluation System of Territorial and Spatial Planning

As mentioned earlier, the Dual-evaluation of territorial and spatial planning is the premise and foundation for the preparation of territorial and spatial planning, and it is also a work that needs to be implemented at the national, provincial, municipal, and county levels. Therefore, an appropriate means is needed to ensure the scientific, standardized, and effective implementation of Dual-evaluation. However, due to the complexity of the Dual-evaluation computation work, the existing GC pattern, whether manual mode or programming mode, cannot meet the needs of the Dual-evaluation work. In view of the above situation, the author developed the "Intelligent data processing and analysis system for Dual-evaluation of territorial and spatial planning" (hereinafter referred to as the "Dual-evaluation system") according to the actual needs of the Dual-evaluation work during the trial evaluation in Guangzhou. The biggest difference with the traditional GIS application system is that the system is constructed according to the new GC pattern and uses G language to develop the evaluation function. It can be said that the Dual-evaluation system is the first application system constructed by G language. The author hopes that through the demonstration of this system, G language will be popularized and applied to other fields related to geographical analysis, such as environmental evaluation, site selection analysis, etc.

6.1 System Development

6.1.1 Overall Structure of the System

The Dual-evaluation system is built in the G language integrated development environment, and the overall structure is shown in Fig. 6.1.

The system has a three-tier structure, namely, data layer, functional service layer and user layer.

The data layer includes all kinds of spatial data used for Dual-evaluation, involving many professional data contents such as ecology, land resources, water resources,

© Surveying and Mapping Press 2021

W. Zhou, *A New GeoComputation Pattern and Its Application in Dual-Evaluation*,
https://doi.org/10.1007/978-981-33-6432-5_6

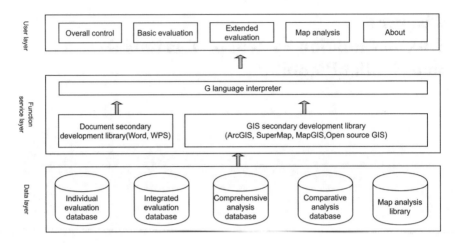

Fig. 6.1 Overall structure of Dual-evaluation system

environment, disasters, climate, and location, specifically including individual evaluation database, integrated evaluation database, comprehensive analysis database, comparative analysis database and map analysis database.

The function service layer is mainly G language interpreter, which is responsible for analyzing and processing intelligent documents. The operation of G language interpreter needs the support of document secondary development library and GIS secondary development library. Document secondary development library refers to MS Word or Kingsoft WPS secondary development library, and GIS secondary development library refers to the secondary development library provided by ArcGIS, SuperMap, MapGIS and open source GIS (currently G language interpreter only supports ArcGIS and will support the other three GIS platforms in the future).

The user layer provides various evaluation functions of Dual-evaluation for Dual-evaluation users, mainly including basic evaluation (e.g. individual evaluation, integrated evaluation, and comprehensive analysis), extended evaluation (e.g. optional evaluation and comparative analysis) and map analysis (e.g. visual analysis, mapping template and map analysis).

6.1.2 System Function Development

1. Overview

Different from the development of traditional GIS application system, the development of Dual-evaluation system is unique, mainly reflected in three aspects:

(1) The development and operation environment of the Dual-evaluation system is MS Word or Kingsoft WPS, which is familiar to the public.
(2) The non-evaluation functions of the Dual-evaluation system, such as environment setting, document maintenance, visual analysis, template customization, etc., need not be developed, and can directly use the functions in G language IDE.
(3) The development of the evaluation functions of the Dual-evaluation system (such as individual evaluation, integrated evaluation and comprehensive evaluation, etc.) does not need general programming languages (such as C # and Python), but only requires users to use G language to describe the evaluation process on the basis of fully understanding and being familiar with each evaluation model.

2. Development of evaluation function

The development of evaluation function is mainly to complete the compiling of intelligent documents according to the contents of Sect. 5.3, as follows.

(1) Basic parameter table

According to the requirements of the Dual-evaluation task, complete the filling of the [Basic parameter table] (as shown in Fig. 6.2).

(2) Task setting table

According to the requirements of *Technical Guidelines*, analyze and sort out the Dual-evaluation GC tasks, and complete the filling of the task setting table (as shown in Fig. 6.3).

Serial NO.	Contents	Working path and file name	Description
1.	[Basic workspace]	T20200611Z	Basic working directory
2.	[Range layer]	BaseMap/MapStat.shp	Used to set the working range, coordinate system and extract output map
3.	[Grid size]	20	Used to set the grid size in meters
4.	[Statistical layer]	BaseMap/MapStat.shp	For partition statistics
5.	[Thematic map template]	Atlas1.mxd	For making thematic maps
6.	[Atlas template]	ZPoint.mxd	For making questionable point map
7.	[Evaluation area]	**city	Outputi n statistical report
8.	[Export space]	T20200611Z	Basic workspace backup directory (support absolute path and relative path)

Fig. 6.2 [Basic parameter table] of Dual-evaluation system

Serial NO.	Task name	Calculate or not	Workspace	Table location (input and control table)
1.	Individual evaluation_ecology		DX/DX1	Table 7
2.	Individual evaluation_land resource		DX/DX2	Table 25
3.	Individual evaluation_water resource	Y	DX/DX3	Table 35
4.	Individual evaluation_climate	Y	DX/DX4	Table 41
5.	Individual evaluation_environment	Y	DX/DX5	Table 47
6.	Individual evaluation_disaster	Y	DX/DX6	Table 57
7.	Individual evaluation_comprehensive advantage (provincial level)		DX/DX71	Table 61
8.	Individual evaluation_Comprehensive advantage (city and county)	Y	DX/DX72	Table 65
9.	Integrated evaluation_ ecological protection	Y	JC/JC1	Table 73
10.	Integrated evaluation_agricultural production	Y	JC/JC2	Table 77
11.	Integrated evaluation_urban construction	Y	JC/JC3	Table 81
12.	Integrated evaluation_agricultural scale	Y	JC/JC4	Table 85
13.	Integrated evaluation_urban scale	Y	JC/JC5	Table 89
14.	Comprehensive analysis_resource endowment		ZH/ ZH1	Table 93
15.	Comprehensive analysis_ problem risk		ZH/ZH2	Table 98
16.	Comprehensive analysis_ agricultural potential		ZH/ ZH3	Table 101
17.	Comprehensive analysis_ urban potential		ZH/ ZH4	Table 107
18.	Comprehensive analysis_ scenario analysis		ZH/ ZH5	Table 113
19.	Optional evaluation_ marine development		KX/KX1	Table 123
20.	Optional evaluation_ cultural protection		KX/KX2	Table 127
21.	Optional evaluation_ mineral resources		KX/KX3	Table 131
22.	Comparative analysis_ status comparison		DB/DB1	Table 135
23.	Comparative analysis_ three-lines comparison		DB/DB2	table 146
24.	Comparative analysis_ comparison of province and city		DB/DB3	table 156
25.	Comparative analysis_ temporary comparison		DB/DB4	table 164
26.	Map analysis_ thematic Atlas		DT/DT1	table 168
27.	Map analysis_ questionable point map		DT/DT2	table 170
28.	Map analysis_ evaluation results		DT/DT3	table 173

Fig. 6.3 Dual-evaluation system task setting table

(3)   Description of GC task

According to Sect. 4.2.3, 28 GC tasks are described in G language. In the third part of this book will introduce the main GC tasks in detail. Through these examples, readers can not only be familiar with the grammar rules of G language, but also create their own GC tasks through G language.

(4)   Configuration of GC tasks

Edit the menu file "Menu.txt" and configure 28 GC tasks to the function menu in "G Language IDE". The specific configuration scheme is shown in Table 6.1.

Table 6.1   GC task configuration table

Sequence no.	GC task name	Functions	Module
1.	Individual evaluation_ecology	Individual evaluation	Basic evaluation
2.	Individual evaluation_land resource		
3.	Individual evaluation_water resource		
4.	Individual evaluation_climate		
5.	Individual evaluation_environment		
6.	Individual evaluation_disaster		
7.	Individual evaluation_comprehensive advantage (provincial level)		
8.	Individual evaluation_comprehensive advantage (city and county)		
9.	Integrated evaluation_ecological protection	Integrated evaluation	
10.	Integrated evaluation_agricultural production		
11.	Integrated evaluation_urban construction		
12.	Integrated evaluation_agricultural scale		
13.	Integrated evaluation_urban scale		
14.	Comprehensive analysis_resource endowment	Comprehensive analysis	
15.	Comprehensive analysis_problem risk		
16.	Comprehensive analysis_agricultural potential		

(continued)

Table 6.1 (continued)

Sequence no.	GC task name	Functions	Module
17.	Comprehensive analysis_urban potential		
18.	Comprehensive analysis_scenario analysis		
19.	Optional evaluation_marine development	Optional evaluation	Extended evaluation
20.	Optional evaluation_cultural protection		
21.	Optional evaluation_mineral resources		
22.	Comparative analysis_status comparison	Comparative analysis	
23.	Comparative analysis_three-lines comparison		
24.	Comparative analysis_comparison of province and city		
25.	Comparative analysis_temporary comparison		
26.	Map analysis_thematic atlas	Map analysis	Map analysis
27.	Map analysis_questionable point map		
28.	Map analysis_evaluation results		

(5) Debugging GC task

Each GC task is debugged through a single-step operation mode or a multi-step operation mode, and the intermediate GC results can be observed through the "Visualization" function in the [Map analysis] module to judge the correctness of the GC results.

6.1.3 Basic Database Construction

The basic database is the basis of the Dual-evaluation work. Therefore, it is necessary to collect, sort out and process relevant data according to the requirements of *Technical Guidelines*. When collecting data, the authority, accuracy, and effectiveness of the data should be ensured. The 2000 National Geodetic Coordinate System

(CGCS2000) and Gauss-Kruger projection should be uniformly adopted in data processing.

For the Dual-evaluation system, considering the flexibility of the Dual-evaluation work and the convenience of data exchange, the file system is adopted to organize the spatial database. Specifically, it includes two aspects.

1. Data format

The spatial data to be processed by the Dual-evaluation system includes vector and raster data formats. The vector data adopts shapefile format and the raster data adopts raster format of ArcGIS.

2. Data organization

For the organization of spatial data, the file management mode is adopted, and the data involved in different GC tasks (including original data, intermediate GC results and final GC results) are stored in the same file directory, to facilitate the management and use of processing results. Similar GC tasks can be managed in the same GC task group directory. Figure 6.4 shows the directory structure of the spatial data organization of the Dual-evaluation system.

(1) Directory naming rules

In order to facilitate the search and use of data, for directory naming, the first two phonetic initials are used in the task category directory, while the task category abbreviation + serial number are used to name the specific task category

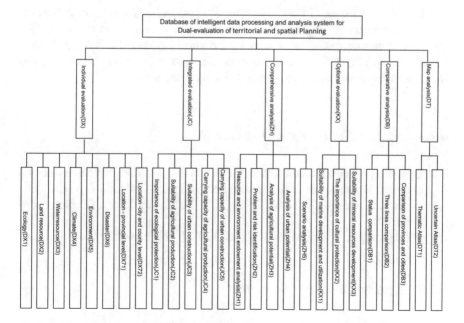

Fig. 6.4 Directory structure of spatial data organization of Dual-evaluation system

directory. For example, the directory tree in Fig. 6.4 is named as individual evaluation-DX, and its sub-directories are named as ecological evaluation-DX1, land resource evaluation-DX2, water resource evaluation-DX3, climate evaluation-DX4, environment evaluation-DX5, disaster evaluation-DX6, location evaluation (province)-DX71 and location evaluation (city and county)-DX72.

(2) Layer naming rules

In order to facilitate the data analysis in the evaluation process, all layers in the evaluation process are uniformly named except the above directory and directory naming rules. Each layer has two names, namely, the layer name and the physical layer name. The layer name is used for business communication and uses professional terminology, such as [Agricultural farming conditions] and [Agricultural water supply conditions]. The physical layer name is used for the operation and internal management of the actual layer. The physical layer adopts the naming method of "directory name _ pinyin abbreviation", such as [Agricultural farming conditions] and [Agricultural water supply conditions], which are named as JC2_NYGZTJ and JC2_NYGSTJ, while the layer in the GC process is named as "directory name _ pinyin abbreviation + serial number", such as [Agricultural carrying capacity 1], which is named as JC2_NYGZTJ1.

6.2   System Installation

The installation process of the Dual-evaluation system is quite special. It does not need to be installed through Setup program. Users only need to set the basic operating parameters according to the contents introduced in Sect. 6.2. After that, when performing specific GC tasks, the system will automatically install the external function library required by the system and register the users.

6.2.1   System Operating Environment

(1) Operating system: Window system (WIN7, WIN8, WIN10), Net framework 4.0 or above (if not, can be installed with Setup directory installation program);
(2) Document system: Microsoft Word 2011–2016, Word 365, Recommended Word 2013;
(3) GIS system: ArcGIS 10.1–10.8, ArcGIS 10.2 is recommended.

6.2.2   System Directory Structure

The directory structure of the system is shown in Fig. 6.5, and the explanation is shown in Table 6.2.

6.2.3 Environment Setting and Registration

1. Set up the custom toolbar

In the Word environment, open the Word option dialog box (as shown in Fig. 6.6) through [File], [Option], [Custom ribbon], [Import/Export] and [Import custom file]. In this dialog box, open the file dialog box (as shown in Fig. 6.7) through the [Import/Export] function.

Navigate to the DAS system installation directory in the file dialog box, and select the "SPJ0414. ExportedUI" file in the [Setup] subdirectory under the directory (note: the file names published by different DAS versions are different, and the latest file is sufficient). At this time, the [G language IDE] shown in Fig. 6.8 will appear in the Word toolbar.

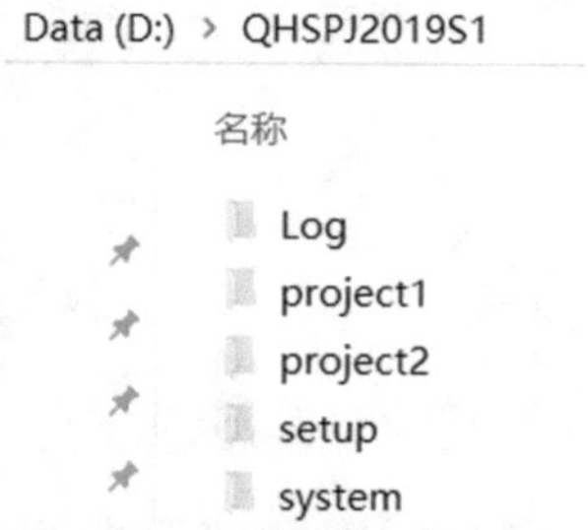

Data (D:) ＞ QHSPJ2019S1

名称

Log

project1

project2

setup

system

Fig. 6.5 Directory structure of the system

Table 6.2 System directory structure description

Sequence No.	Directory	Description
1	Setup	The external support library and customized menu file in MS Word environment are needed for system operation
2	System	The executable file of system operation and its supporting auxiliary files
3	Project1	Das 2019 demonstration system
4	Project2	Urban heat island effect analysis demonstration system
5	Log	Process record of system operation

Fig. 6.6 The first step to set the customized toolbar of Dual-evaluation function group

Fig. 6.7 The second step to set the customized toolbar of Dual-evaluation function group

Fig. 6.8 Dual-evaluation system toolbar

Fig. 6.9 Set system operating environment parameter

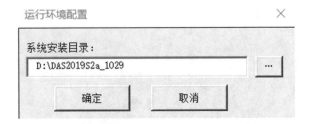

2. **Set the operating environment parameters**

In MS Word environment, click [Environment settings] in [Overall control] in the [G language IDE] toolbar, and the [Running environment configuration] dialog box shown in Fig. 6.9 will appear.

Set the System Installation Directory in the dialog box, which is the path where DAS system packages are decompressed. When running for the first time, the system will automatically fill in the content according to the path where the MS Word document is located. Of course, users can also set the system directory to other paths through this function.

3. **System registration**

After the system directory is set up, users can run [Land resources evaluation] module through the following steps, [G language IDE], [Basic evaluation] and [Single element] (as shown in Fig. 6.10).

Before registering, the system will check whether the necessary basic library pywin is installed. If it is not installed, it will be installed automatically, as shown in Fig. 6.11.

After the installation of pywin, the basic function library, the system will prompt to close the running window and run [Land resource evaluation] again (as shown in Fig. 6.12).

If the user has not registered before, the system will give a prompt as shown in Fig. 6.13. At this time, close the running window again and run [Land resource evaluation] again.

After running [Land resources evaluation] again, the prompt information shown in Fig. 6.14 will appear. At this time, the user can send short messages as required.

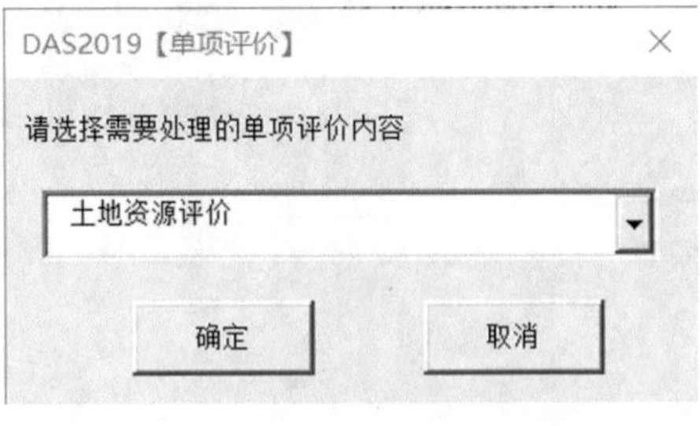

Fig. 6.10 Individual evaluation dialog

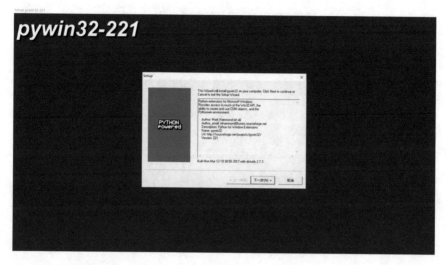

Fig. 6.11   Install the basic functionality library pywin

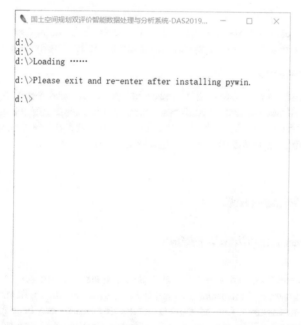

Fig. 6.12   Prompt information after installation of basic function library pywin

国土空间规划双评价智能数据处理与分析系统-DAS2019
d:\>
d:\>DAS2019 (Sld) Loading ……
d:\>The system is not registered, please exit and re-enter.
d:\>

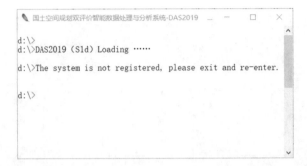

Fig. 6.13 Prompt for system registration

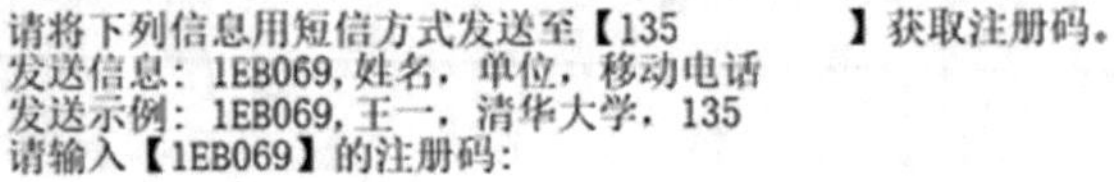

请将下列信息用短信方式发送至【135 】获取注册码。
发送信息：1EB069,姓名，单位，移动电话
发送示例：1EB069,王一，清华大学，135
请输入【1EB069】的注册码：

Fig. 6.14 System registration information

After that, the user will receive the registration code information, and the user can complete the registration of the system after inputting the information.

4. **System operation**

After completing the registration, the user can perform [land resource evaluation]. The system will automatically install other external libraries during the first execution (as shown in Fig. 6.15). After the installation is completed, the system will enter the normal operation state (as shown in Fig. 6.16).

6.3 System Instructions

6.3.1 System Application Mode

"Dual-evaluation" system is a GIS application system based on a new concept. Compared with other traditional GIS application systems, its application mode has its own unique features. One is that the evaluation model can be adjusted in MS Word or Kingsoft WPS documents; the other is that the GC results are directly embedded in MS Word or Kingsoft WPS documents, which can be used for the preparation of "Dual-evaluation" report. The application of "Dual-evaluation" system to "Dual-evaluation" includes the following four steps.

(1) The "Dual-evaluation" personnel adjust the evaluation process and model parameters in the intelligent MS Word or Kingsoft WPS document according to the *Technical Guidelines* and the characteristics of the evaluation area (if

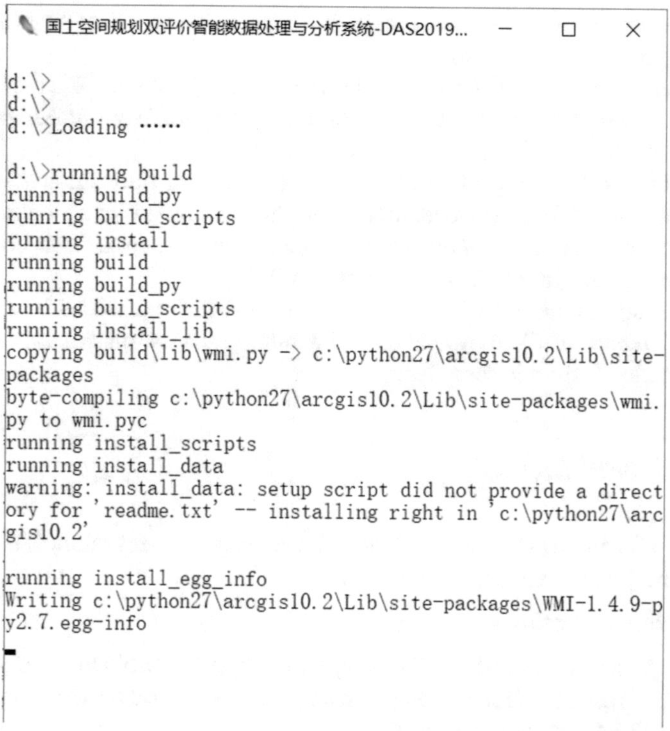

Fig. 6.15 Automatic installation of external library for Dual-evaluation system

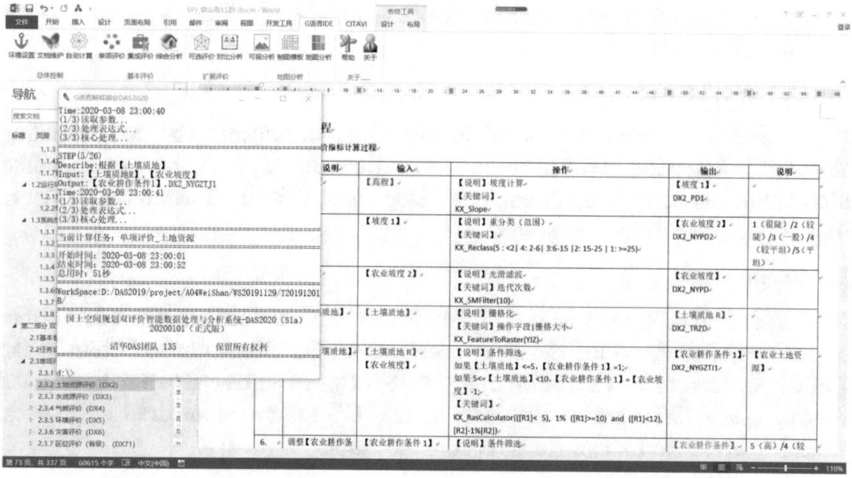

Fig. 6.16 Operation interface of Dual-evaluation system

there are no special requirements, it is not necessary to adjust the MS Word or Kingsoft WPS documents).

(2) Connect the data in the data list or the national and provincial "Dual-evaluation" results with the data described in the intelligent MS Word or Kingsoft WPS document.

(3) In MS Word or Kingsoft WPS, single task mode or fully automatic mode is adopted to perform task operation. After the operation is completed, a "Dual-evaluation basic chart report" containing all thematic maps and statistical tables can be formed in MS Word or Kingsoft WPS.

(4) According to the thematic maps and statistical tables provided by the "Basic chart report of Dual-evaluation" and other relevant materials, the "Dual-evaluation" report is compiled.

6.3.2   Overall Control

The [Overall control] function module includes [Environment setting], [Document maintenance] and [Automatic computation].

1. **Environment settings**

It is used to set the installation directory of the system. Different versions of the system can be used in the same computer, and can be switched through this function. See Sect. 6.2.3 for specific operations.

2. **Document maintenance**

It mainly includes four sub-functions: [Document refresh], [Export space], [Empty chart] and [Label check].

(1) Document refresh

When tables are added or deleted in intelligent documents, the corresponding "domain" information needs to be updated in time. At this time, the corresponding information can be updated through the [Document refresh] function (as shown in Fig. 6.17) and the file can be saved.

(2) Export space

This function exports the basic input data (excluding intermediate result data) from the basic workspace to the directory specified by the "Export space" item in the [Basic parameter table] (as shown in Fig. 6.18). This function can make the original working space "Slimmed down" from several G's to several hundred megabytes, which is convenient for storage and communication.

(3) Empty chart

This function is used to delete thematic maps and statistical table data generated by computation in intelligent documents to facilitate the next computation (as shown in Fig. 6.19).

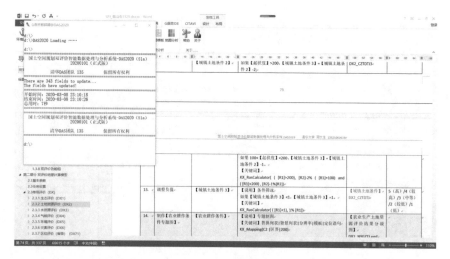

Fig. 6.17 Example of the [Document refresh] process

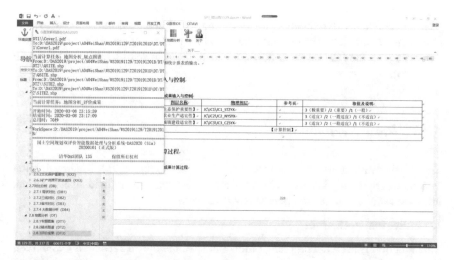

Fig. 6.18 Example of the [Export space] process

(4) ID inspection

This function is used to check whether the ID of each GC task identified in the [Task table] (as shown in Fig. 6.20) is consistent with the ID in the [Input and control table] (as shown in Fig. 6.21) (as shown in Fig. 6.22).

3. Automatic computation

The computation of each evaluation model in the Dual-evaluation system can adopt two computation modes, single computation mode and automatic computation mode. A single computation mode can execute a single GC task separately through the

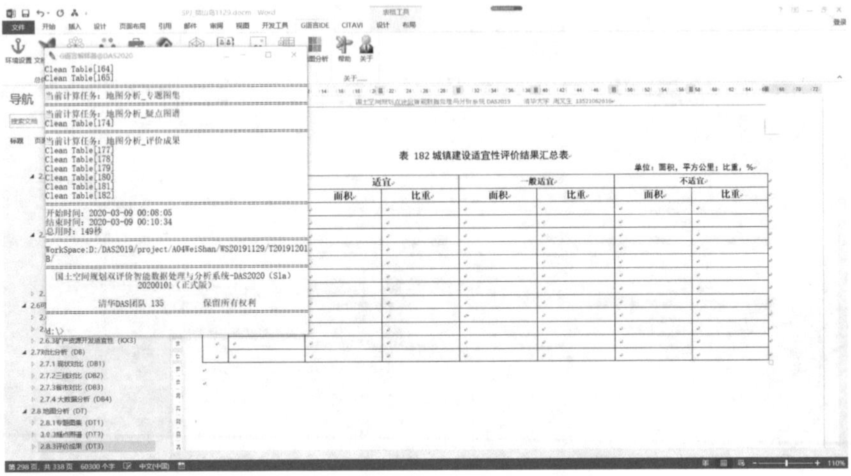

Fig. 6.19 Example of the [Empty chart] process

[Basic evaluation], [Extended evaluation] and [Map analysis] modules, while the [Automatic computation] mode automatically executes the selected GC task through the settings in the [Calculate or not] column in the task setting table (as shown in Fig. 6.23).

6.3.3 Basic Evaluation

The [Basic evaluation] function module is the core content of the Dual-evaluation system, which mainly includes 17 evaluations in 3 categories: individual evaluation, integrated evaluation, and comprehensive evaluation.

1. **Individual evaluation**

This function is used to complete 7 individual evaluations in *Technical Guidelines*, namely ecological evaluation, land resource evaluation, water resource evaluation, climate evaluation, environment evaluation, disaster evaluation and comprehensive advantage evaluation. The comprehensive advantage evaluation is divided into comprehensive advantage evaluation (provincial level) and comprehensive advantage evaluation (city and county). See Chap. 7 for details.

2. **Integrated evaluation**

This function is used to complete the five integrated evaluations in *Technical Guidelines*, namely, the importance of ecological protection, the suitability of agricultural production, the suitability of urban production, the carrying scale of agricultural production and the carrying scale of urban construction. See Chap. 8 for details.

Serial NO.	Task name	Calculate or not	Workspace	Task ID
1.	Individual evaluation_ ecology		DX/DX1	table 7
2.	Individual evaluation_ land resource		DX/DX2	table 25
3.	Individual evaluation_ water resource	Y	DX/DX3	table 35
4.	Individual evaluation_ climate	Y	DX/DX4	table 41
5.	Individual evaluation_ environment	Y	DX/DX5	table 47
6.	Individual evaluation_disaster	Y	DX/DX6	table 57
7.	Individual evaluation_comprehensive advantage (provincial level)		DX/DX71	table 61
8.	Individual evaluation_ comprehensive advantage (city and county)	Y	DX/DX72	table 65
9.	Integrated evaluation_ ecological protection	Y	JC/JC1	table 73
10.	Integrated evaluation_ agricultural production	Y	JC/JC2	table 77
11.	Integrated evaluation_ urban construction	Y	JC/JC3	table 81
12.	Integrated evaluation_ agricultural scale	Y	JC/JC4	table 85
13.	Integrated evaluation_ urban scale	Y	JC/JC5	table 89

Fig. 6.20 Task ID in [Task setting] table

3. Comprehensive evaluation

This function is used to complete the five comprehensive evaluations in *Technical Guidelines*, namely, resource and environment endowment analysis, problem and risk identification, agricultural potential analysis, urban potential analysis and scenario analysis.

6.3.4   Extended Evaluation

Extended evaluation is an optional evaluation content, and each evaluation area needs to decide on its own according to the characteristics of its own area and the needs of analysis. It mainly includes 7 evaluations in two categories: optional evaluation and comparative analysis.

Table 7 Input and control of ecological evaluation

Serial NO.	Layer semantics	Layer name	Reference page	Value and description
1.	[Ecosystem service function quantity]	DX1_FWGNL.shp		SL - Number of species
2.	[Important drinking water source]	DX1_YYSYD.shp		DJ-1 (Important drinking water source)
3.	[Key Biodiversity Conservation Areas]	DX1_BHGJQ.shp		DJ-1 (Key protection areas)
4.	[Sensitivity of soil erosion]	DX1_STLS.shp		DJ - Level 3 (Very sensitive) / 2 (Sensitive) / 1 (General sensitive)
5.	[Desertification sensitivity]	DX1_SMH.shp		DJ - Level 3 (Very sensitive) / 2 (Sensitive) / 1 (General sensitive)
6.	[Sensitivity of rocky desertification]	DX1_SHMH.shp		DJ - Level 3 (Very sensitive) / 2 (Sensitive) / 1 (General sensitive)
7.	[Coastal erosion sensitivity]	DX1_HAQS.shp		DJ - Level 3 (Very sensitive) / 2 (Sensitive) / 1 (General sensitive)
8.	[Soil texture]	DX\DX2\DX2_TRZD.shp		ZD- Soil texture
9.	[Meteorological]	DX\DX4\DX4_QX.shp		JYL- Precipitation ZFL- Evaporation QSL- Precipitation erosivity
10.	[Vegetation cover]	DX1_NDVI1		
11.	[Height]	DX\DX2\DX2_GC5		

Fig. 6.21 Task ID in [Input and control table]

Fig. 6.22 Example of [ID check] process

Serial NO.	Task name	Calculate or not	Workspace	Task ID
1.	Individual evaluation_ ecology		DX/DX1	table 7
2.	Individual evaluation_ land resource		DX/DX2	table 25
3.	Individual evaluation_ water resource	Y	DX/DX3	table 35
4.	Individual evaluation_ climate	Y	DX/DX4	table 41
5.	Individual evaluation_ environment	Y	DX/DX5	table 47
6.	Individual evaluation_disaster	Y	DX/DX6	table 57
7.	Individual evaluation_comprehensive advantage (provincial level)		DX/DX71	table 61
8.	Individual evaluation_ comprehensive advantage (city and county)	Y	DX/DX72	table 65
9.	Integrated evaluation_ ecological protection	Y	JC/JC1	table 73
10.	Integrated evaluation_ agricultural production	Y	JC/JC2	table 77
11.	Integrated evaluation_ urban construction	Y	JC/JC3	table 81
12.	Integrated evaluation_ agricultural scale	Y	JC/JC4	table 85
13.	Integrated evaluation_ urban scale	Y	JC/JC5	table 89

Fig. 6.23 Example of settings in task settings table in automatic computation

Optional evaluations mainly include the suitability evaluation of marine development and utilization, the importance evaluation of cultural protection and the suitability evaluation of mineral resources development referred to in *Technical Guidelines*. See Chap. 9 for details.

Comparative analysis is mainly used for comparative analysis of various evaluations, mainly including current situation comparative analysis, three-lines comparative analysis, provincial and municipal comparative analysis, and temporary comparative analysis. Temporary comparative analysis can build a comparative analysis model as required to facilitate users to understand the evaluation results. See Chap. 10 for details.

6.3.5 Map Analysis

It is mainly used to output various thematic maps and reports in Dual-evaluation, including visual analysis, mapping template and map analysis.

1. Visual analysis

It is used to visually edit and analyze various evaluation-related spatial data in ArcMap environment. It includes individual evaluation spatial data, integrated evaluation spatial data, comprehensive analysis spatial data and optional evaluation spatial data. During the specific operation, the user can drag the intermediate result or result of the computation from Catalog into ArcMap for visual analysis (as shown in Fig. 6.25) through the layer information in the "Output" column in the [GC process table] (as shown in Fig. 6.24) as required.

2. Mapping templates

Templates used to edit various thematic map outputs, including editing, and managing thematic map templates and questionable point map templates.

(1) Thematic map

Template Used to edit and manage thematic map templates. In ArcMap environment, users can edit the name, compilation unit, north compass, scale, and legend of thematic map (as shown in Fig. 6.26).

In addition, the important function of this function is to edit and manage the style of thematic map output layer. See Sect. 4.4.5 for the use of layer styles.

(2) Questionable point map template

It is used for editing and managing questionable point map templates. This operation is also carried out in ArcMap environment (as shown in Fig. 6.27). Unlike thematic

Table 8 Calculation process of ecological evaluation index

Step	Operation instruction	Input	Operation	Output	Description
1.	Output precipitation raster	[Range layer]	[Description] Modification [Keywords] Number or fireld KX_MODIFYRASTER(607.4)	[Water conservation 1] DX1_SYHY1	Importance of water conservation function (pricipitaton) 607.4
2.	Output evaporation raster	[Range layer]	[Description] Mdifificatin [Keywords] Number or field KX_MODIFYRASTER1019.0)	[Water conservation 2] DX1_SYHY2	Importance of water conservation function (evaporation) 1019.0
3.	Water conservation calculation	[Water conservation 1] [Water conservation 2] [Water conservation 3]	[Description] Reclassification (class modification) [Keywords]New class: old class 1, old class 2,...... KX_Reclass_Vec(DLBM,100#2.67:0301,0302,0303,0304,0307|4.26:0305,0606|8.2:0401,0402,0403,0404|9.57:0201,0203,0204|7.9:0202|34.7:0101|49.69:0102,0103)	[Water conservation 3] DX1_SYHY3	Surface runoff coefficient See p17
4.	Classification	[Water conservation 4]	[Description] Raster calculator [Keywords] KX_RasCalculator([R1]-[R1]*[R3]-[R2])	[Water conservation 4] DX1_SYHY4	
5.	Clip [Water conservation 5]	[Water conservation 5] [Range layer]	[Description] Reclassification (equidistant) [Keywords] Class 1|class 2|...... KX_Reclass(1|2|3)	[Water conservation 5] DX1_SYHY5	
6.	Clip [Water conservation 5]	[Water conservation 5] [Range layer]	[Description]Clip [Keywords] Clipped layer,clipping layer KX_ExtractByMask	[Water conservation] DX1_SYHY	3 (Extremely important) / 2 (Important) / 1 (Generally important)

Fig. 6.24 Middle GC result information in GC process table

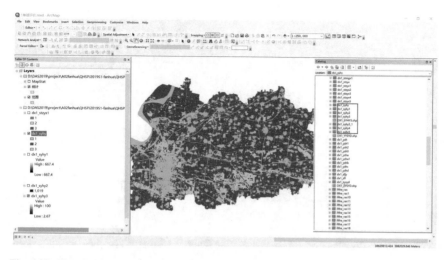

Fig. 6.25 Visualization of intermediate GC results in ArcMap

map templates, questionable point map templates are mainly used for editing and managing layer styles.

The use of layer styles in the questionable point map template is basically the same as that in the thematic map template, except that the keyword "KX_Mapping" is selected differently in the "Template" parameter. The thematic map template is "1" and the questionable point map template is "2" (as shown in Fig. 6.28).

Figure 6.29 is an example of the application of questionable point map template layer style.

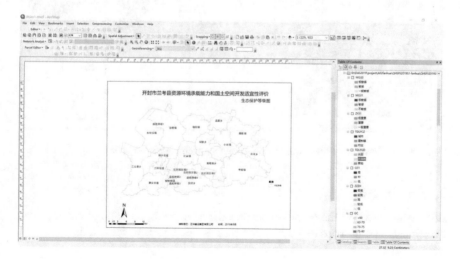

Fig. 6.26 Edit thematic map template

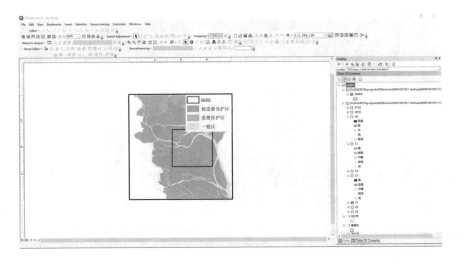

Fig. 6.27 Edit questionable point map template

8.	制作#	【疑点】	【说明】专题制图	7	【灾害危险性评价
		【灾害危险性】	【关键词】替换列表|背景列表|分辨率|模板|定位语句		图】
			KX_Mapping (MARK,G4_3| |200|2|*)		DT2_DT8. emf

Fig. 6.28 Example of application of questionable point map template layer style

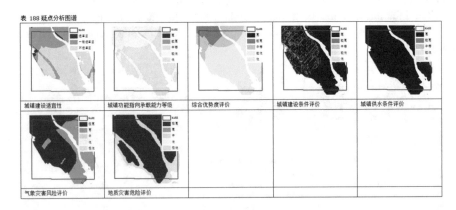

Fig. 6.29 Application results of questionable point map template layer style

3. Map analysis

It is mainly used for the output of various thematic maps, including thematic atlas, questionable point maps and evaluation results. See Chap. 11 for details.

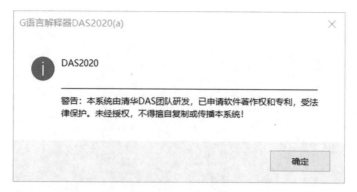

Fig. 6.30 Copyright information of Dual-evaluation system

6.3.6 About

It is used to browse the contents of *Technical Guidelines* and display the copyright information of the system at any time during the Dual-evaluation work, as shown in Fig. 6.30.

6.3.7 Problems and Solutions

There will be various problems during the installation and use of DAS2019. Appendix II has listed the common problems and solutions. For other problems, readers can pay attention to the WeChat official account "SPJ_DAS2019".

Part III
The Dual-Evaluation GC Models

This part is the practice of the new GC pattern, which mainly introduces the evaluation models of Dual-evaluation of territorial and spatial planning compiled by G language. From a programming point of view, these evaluation models are actually program codes compiled in G language in MS Word or KingSoft WPS documents, but these codes can be read and used by technicians without programming experience. It is hoped that this part of the content can help readers master G language and play a role in the Dual-evaluation work. Of course, readers can also build their own evaluation models through this part of learning, such as planning evaluation models, environmental evaluation models, etc. In addition, there are several points that need to be explained.

(1) The models introduced in the book are basically compiled according to the contents *Technical Guidelines (June Edition)*. Users can adjust the models according to the latest *Technical Guidelines* when using it.

(2) When users carry out Dual-evaluation with reference to this book, they can adjust these evaluation models according to the collected data and the objective conditions of the evaluation area.

(3) The models introduced in this book basically cover all the contents required for dual-evaluation. Individual evaluation models not introduced, such as "bearing scale", "potential analysis" and "scenario analysis", can be compiled by themselves according to the *Technical Guidelines* or applied on the WeChat official account at the bottom.

(4) The "Evaluation Method" part of the GC task only briefly introduces the relevant evaluation models. For details, please refer to the relevant instructions in *Technical Guidelines (June Edition)* according to the page numbers given.

(5) The content of the "Thematic map-statistical table" part of the GC task is only schematic, and the author only shows representative evaluation results in individual GC tasks.

(6) The evaluation model introduced in this book needs to run with the support of the G language interpreter. Users can scan the two-dimensional code of this page and apply for the G language interpreter on the official account of WeChat.

Two-dimensional code for Dual-evaluation of DAS official account

Chapter 7
Individual Evaluation

Individual evaluation is the basic evaluation content of the Dual-evaluation work, including 7 evaluation contents such as ecology, land resource, water resource, climate, environment, disaster, and location. Among them, the evaluation methods of location evaluation for provincial and city and county are different.

7.1 Ecological Evaluation

7.1.1 Evaluation Method

This model is referred to pages 14–26 in *Technical Guidelines (June Edition)*.

Due to the complexity of ecological evaluation, if the introduction of all evaluation processes takes up a lot of space, the introduction of ecological service function importance evaluation and ecological sensitivity evaluation here only takes the intermediate evaluation results as input. At the same time, to let readers to know the more detailed evaluation process, the relatively independent salinization sensitivity evaluation is introduced in detail. There is a complete model in the official version of the Dual-evaluation software introduced in the book.

1. **Evaluation methodology**

$$[\text{Ecosystem service importance}] = f([\text{Ecosystem service quantity}],$$
$$[\text{Important drinking water source}], [\text{Biodiversity conservation key area}]) \quad (7.1)$$

$$[\text{Ecosystem service function}] = \text{Max}([\text{Importance of biodiversity conservation}],$$
$$[\text{Importance of water conservation}], [\text{Importance of soil and}$$
$$\text{water conservation}], [\text{Importance of windbreak and sand fixation}],$$
$$[\text{Importance of coastal protection}]) \quad (7.2)$$

© Surveying and Mapping Press 2021
W. Zhou, *A New GeoComputation Pattern and Its Application in Dual-Evaluation*,
https://doi.org/10.1007/978-981-33-6432-5_7

$$[\text{Ecological sensitivity}] = f([\text{Soil erosion sensitivity}],$$
$$[\text{Rocky desertification sensitivity}], [\text{Desertification sensitivity}]$$
$$[\text{Coastal erosion and sand source loss sensitivity}]) \qquad (7.3)$$

$$[\text{Salinization sensitivity}] = f([\text{Evaporation/Precipitation}], [\text{Groundwater}$$
$$\text{mineralization}], [\text{Groundwater depth}], [\text{Soil texture}]) \qquad (7.4)$$

2. Evaluation steps

The evaluation process is shown in Fig. 7.1.

(1) Take the maximum values of [Biodiversity conservation importance], [Water conservation Importance], [Soil and water conservation importance], [Windbreak and sand fixation importance] and [Coastal protection importance] as [Ecosystem service function quantity].
(2) Use [Important drinking water source] to modify [Ecosystem service function quantity] and obtain [Importance of ecosystem service function 1].
(3) Modify [Importance of ecosystem service function 1] with [Key area of biodiversity conservation] to obtain the [Importance of ecosystem service function].
(4) Take the maximum values of [Soil erosion Sensitivity], [Rocky Desertification Sensitivity], [Desertification Sensitivity], [Coastal Erosion and Sand Source Loss Sensitivity] as [Ecological sensitivity].
(5) Calculate [Evaporation/Precipitation], [Groundwater mineralization], [Groundwater depth] and [Soil texture] according to the following formula to obtain [Salinization sensitivity 1].

$$[\text{Salinization sensitivity}] = \sqrt[4]{I \times M \times D \times K} \qquad (7.5)$$

In the formula, I, M, D and K are the sensitivity classification values of evaporation/rainfall, groundwater salinity, groundwater depth and soil texture factors in the evaluation area respectively. See Table 7.1 for the assignment of each factor.

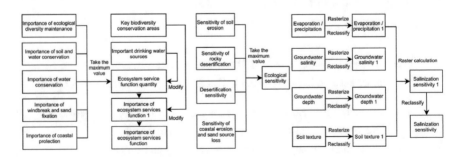

Fig. 7.1 GC flow chart of ecological evaluation

Table 7.1 Grading and assignment of salinization sensitivity evaluation factors

Evaluation factors	Higher	High	Medium	Low	Lower
Hierarchical assignment	9	7	5	3	1
Evaporation/Precipitation	≥ 15	10–15	3–10	1–3	<1
Groundwater salinity (g/l)	≥ 25	10–25	5–10	1–5	<1
Groundwater depth (m)	<1	1–3	3–6	6–10	≥ 10
Soil texture	Sandy loam	Loam	Clay loam	Clay	Sandy soil

Table 7.2 Salinization sensitivity grading reference thresholds

Classification of salinization sensitivity	Higher	High	Medium	Low	Lower
Salinization sensitivity	≥ 7.0	6.0–7.0	5.0–6.0	3.0–5.0	<3.0

(6) Reclassify [Salinization sensitivity 1] according to Table 7.2 to obtain [Salinization sensitivity].

7.1.2 Input and Control

Table 7.3 is the [Input and control table] of ecological evaluation.

7.1.3 GC Process

Table 7.4 shows the GC process of ecological evaluation indexes.

7.2 Evaluation of Land Resources

7.2.1 Evaluation Method

This model is referred to *Technical Guidelines* (June Edition), pages 27–28.

1. **Evaluation methodology**

$$[\text{Farming conditions}] = f([\text{Slope}], [\text{Soil texture}], [\text{Waters}]) \tag{7.6}$$

$$[\text{Urban construction conditions}] = f([\text{Slope}], [\text{Elevation}], [\text{Topographic relief}]) \tag{7.7}$$

Table 7.3 [Input and control table] of ecological evaluation

Serial No.	Layer name	Physical layer	Reference page	Value and description
1.	[Importance of biodiversity conservation]	DX1_SHDYX		3(Very important)/2(Important)/1(Generally important)
2.	[Importance of soil and water conservation]	DX1_SHBC		3(Very important)/2(Important)/1(Generally important)
3.	[Importance of water conservation]	DX1_SYHY		3(Very important)/2(Important)/1(Generally important)
4.	[Importance of windbreak and sand fixation]	DX1_FSGS		3(Very important)/2(Important)/1(Generally important)
5.	[Importance of coastal protection]	DX1_HAFH		3(Very important)/2(Important)/1(Generally important)
6.	[Important drinking water sources]	DX1_ZYSYD.shp		Polygon layer
7.	[Key conservation areas of biodiversity]	DX1_BHGJQ.shp		Polygon layer
8.	[Sensitivity of soil erosion]	DX1_STLS		3(Extremely sensitive)/2(Sensitive)/1(General sensitive)
9.	[Sensitivity of rocky desertification]	DX1_SMH		3(Extremely sensitive)/2(Sensitive)/1(General sensitive)
10.	[Desertification sensitivity]	DX1_SMH		3(Extremely sensitive)/2(Sensitive)/1(General sensitive)
11.	[Sensitivity of coastal erosion and sand source loss]	DX1_HAQS		3(Extremely sensitive)/2(Sensitive)/1(General sensitive)
12.	[Evaporation/precipitation]	DX1_ZFJY.shp		Polygon layer, ZFJY
13.	[Groundwater salinity]	DX1_DXSKHD.shp		Polygon layer, KHD((g/l))
14.	[Groundwater depth]	DX1_DXSMS.shp		Polygon layer, MS(m)
15.	[Soil texture]	DX1_TRZD.shp		Polygon layer, LX
	[GC control]			

2. Evaluation steps

The evaluation process is shown in Fig. 7.2.

(1) Use [DEM] to compute the slope and reclassify it according to Table 7.5 to obtain [Agricultural slope classification].

Table 7.4 GC process of ecological evaluation indexes

Step	Operation instruction	Input	Operation	Output	Description
1.	Take the maximum value	[Importance of biodiversity conservation] [Importance of soil and water conservation] [Importance of water conservation] [Importance of windbreak and sand fixation] [Importance of coastal protection]	[Description] Take the maximum value [Keywords] KX_Max	[Ecosystem service function quantity]	3(Very important) 2(Important) 1(Generally important)
2.	Modification operation	[Important drinking water sources] [Ecosystem service function quantity]	[Description] Raster modification [Keywords] Field name or value KX_MODIFYRASTER (3)	[Importance of ecosystem service function 1] DX1_FWGN1	
3.	Modification operation	[Key conservation areas of biodiversity] [Importance of ecosystem service function 1]	[Description] Raster modification [Keywords] Field name or value KX_MODIFYRASTER (3)	[Importance of ecosystem service function 2] DX1_STZYX2	
4.	Clip	[Importance of ecosystem service function 2] [Range layer]	[Description] Clip [Keywords] Clipped layer, Clip layer KX_Clip	[Importance of ecosystem service function] DX1_STZYX	3(Very important) 2(Important) 1(Generally important)

(continued)

Table 7.4 (continued)

Step	Operation instruction	Input	Operation	Output	Description
5.	Take the maximum value	[Sensitivity of soil erosion] [Sensitivity of rocky desertification] [Desertification sensitivity] [Sensitivity of coastal erosion and sand source loss]	[Description] Take the maximum value [Keywords] **KX_Max**	[Ecological sensitivity 1] DX1_STMG1	
6.	Clip	[Ecological sensitivity 1] [Range layer]	[Description] Clip [Keywords] Clipped layer, Clip layer **KX_Clip**	[Ecological sensitivity] DX1_STMG	3(Extremely sensitive) 2(Sensitive) 1(General sensitive)
7.	Rasterization	[Evaporation/Precipitation]	[Description] Rasterization [Keywords] Operation field or value| {Grid size} **KX_FeatureToRaster(ZFJY)**	[Evaporation/Precipitation R] DX1_ZFJY	
8.	Reclassification	[Evaporation/Precipitation R]	[Description] Grid reclassification (by range) [Keywords] N1: $<$V1|N2: V1 − V2|...|Nn: $\geq$Vn **KX_Reclass(1: <1| 3: 1–3| 5: 3–10 |7: 10–15 |9: ≥15)**	[Evaporation/Precipitation 1] DX1_ZFJY1	
9.	Rasterization	[Groundwater salinity]	[Description] Rasterization [Keywords] Operation field or value|{Grid size} **KX_FeatureToRaster(KHD)**	[Groundwater salinity R] DX1_DXSKHD	
10.	Reclassification	[Groundwater salinity R]	[Description] Raster reclassification (numerical range) [Keywords] N1: $<$V1|N2: V1 − V2|...|Nn: $\geq$Vn **KX_Reclass(1: <1| 3: 1–5| 5: 5–10 |7: 10–25 |9: ≥25)**	[Groundwater salinity 1] DX1_DXSKHD1	

(continued)

Table 7.4 (continued)

Step	Operation instruction	Input	Operation	Output	Description
11.	Rasterization	[Groundwater depth]	[Description] Rasterization [Keywords] Operation field or value| {Grid size} **KX_FeatureToRaster(MS)**	[Groundwater depth R] DX1_DXSMS	
12.	Reclassification	[Groundwater depth R]	[Description] Raster reclassification (by range) [Keywords] N1: <V1|N2: V1 − V2|...| Nn: ≥Vn **KX_Reclass(9: <1|7: 1–3| 5: 3–6 |3: 6–10|1: ≥10)**	[Groundwater depth 1] DX1_DXSMS1	
13.	Reclassification	[Soil texture]	[Description] Raster reclassification (vector class adjustment) [Keywords] N1: C1, C2, ...|N2: C3, C4, ...|... **KX_Reclass(LX|0#9:Sandy loam | 7: loam | 5: Clay loam | 3: Clay | 1: sandy soil)**	[Soil texture 1] DX1_TRZD1	
14.	Raster calculation	[Evaporation/Precipitation 1] [Groundwater salinity 1] [Groundwater depth 1] [Soil texture 1]	[Description] Raster calculation [Keywords] Algebraic or logical expression **KX_RasCalculator(Power([R1] * [R2] * [R3] * [R4], 0.25))**	[Salinization 1] DX1_YZH1	
15.	Classification	[Salinization 1]	[Description] Raster reclassification (numerical range) [Keywords] N1: <V1|N2: V1 − V2|...|Nn: ≥Vn **KX_Reclass(1: <3|2: 3–5|3: 5–6|4: 6–7|5: ≥7)**	[Salinization 2] DX1_YZH2	
16.	Clip	[Salinization 2] [Range layer]	[Description] Clip [Keywords] Clipped layer, Clip layer **KX_Clip**	[Salinization] DX1_YZH	

(continued)

Table 7.4 (continued)

Step	Operation instruction	Input	Operation	Output	Description
17.	Make thematic map	[Importance of ecosystem service function]	**[Description]** Make thematic map **[Keywords]** Replace list l Background list lResolution l {Template} l {Drawing range} **KX_Mapping(S1lBoundary l200)**	[Classification of ecosystem service function importance evaluation results] DX1_STZYX.emf	
18.	Insert thematic map	[Classification of ecosystem service function importance evaluation results]	**[Description]** Insert thematic map **[Keywords]** Picture heightl {Delete content} **KX_InsertPic (12)**	1	
19.	Produce statistical table	[Statistical layer] [Importance of ecosystem service function]	**[Description]** Statistical table **[Keywords]** Display field# {Unit parameter} l Counted type listl {Data filtering mode} **KX_Statistic(Namel3,2,1)**	2	
20.	Make thematic map	[Ecological sensitivity]	**[Description]** Make thematic map **[Keywords]** Replace list l Background list lResolution l {Template} l {Drawing range} **KX_Mapping(MG32lBoundary l200)**	[Classification map of ecological sensitivity evaluation results] DX1_STMGX.emf	
21.	Insert thematic map	[Classification map of ecological sensitivity evaluation results]	**[Description]** Insert thematic map **[Keywords]** Picture heightl {Delete content} **KX_InsertPic (13)**	3	
22.	Produce statistical table	[Statistical layer] [Ecological sensitivity]	**[Description]** Statistical table **[Keywords]** Display field# {Unit parameter} l Counted type listl {Data filtering mode} **KX_Statistic(Namel3,2,1)**	4	

(continued)

Table 7.4 (continued)

Step	Operation instruction	Input	Operation	Output	Description
23.	Make thematic map	[Salinization]	**[Description]** Make thematic map **[Keywords]** Replace list | Background list |Resolution | {Template} | {Drawing range} **KX_Mapping(C2|Boundary |200)**	[Classification map of salinization sensitivity evaluation results] **DX1_YZHMGX.emf**	
24.	Insert thematic map	[Classification map of salinization sensitivity evaluation results]	**[Description]** Insert thematic map **[Keywords]** Picture height| {Delete content} **KX_InsertPic (13)**	5	
25.	Produce statistical table	[Statistical layer] [Salinization]	**[Description]** Statistical table **[Keywords]** Display field# {Unit parameter} | Counted type list| {Data filtering mode} **KX_Statistic(Name|3,2,1)**	6	

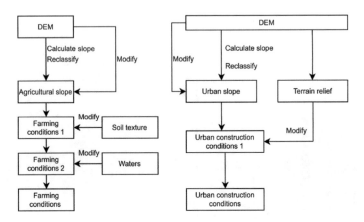

Fig. 7.2   GC flow chart of land resources evaluation

Table 7.5   Agricultural slope grading reference thresholds

Grading of agricultural slope	Flat	Flat slope land	Slope land	Gentle steep slope land	Steep slope land
Slope	≤2	2–6	6–15	15–25	>25

Table 7.6   Reference thresholds for urban slope grading

Grading of urban slope	Flat	Flat slope land	Slope land	Gentle steep slope land	Steep slope land
Slope	≤3	3–8	8–15	15–25	>25

(2) Use [Soil texture] to modify [Agricultural slope classification], and then deducting [Waters] to obtain [Agricultural tillage conditions].

(3) Use [DEM] to compute the slope and reclassify it according to Table 7.6 to obtain [Urban slope classification].

(4) Use [DEM] to modify [Urban slope classification] and obtain the preliminary results of [Urban construction conditions].

(5) Use [DEM] to compute [Terrain relief], and use [Terrain relief] to modify the preliminary results of [Urban construction conditions] to obtain the final [Urban construction conditions].

7.2.2   Input and Control

Table 7.7 shows [Input and control table] of land resources evaluation.

Table 7.7 [Input and control table] of land resource evaluation

Serial No.	Layer name	Physical layer	Reference page	Value and description
1.	[Height]	DX2_GC		Country and province (grid size: 20m * 20m or 30m * 30m), City and county (grid size of 20m * 20m or 30m * 30m)
2.	[Soil texture]	DX2_TRZD		Percentage of silt content in soil
				[GC control]

7.2.3 GC Process

Table 7.8 is the [GC Process table] for land resources evaluation.

7.2.4 Thematic Map and Statistical Table

Figure 7.3 is the first example of thematic map, Fig. 7.4 is the second example of thematic map, Table 7.9 is the statistical table of agricultural farming conditions, and Table 7.10 is the statistical table of urban development conditions.

7.3 Water Resource Evaluation

7.3.1 Evaluation Method

This model is referred to pages 29–31 of the *Technical Guidelines (June Edition)*.

1. **Evaluation methodology**

$$\left[\text{Agricultural water supply conditions}\right] = f(\left[\text{Precipitation}\right]) \qquad (7.8)$$

$$\left[\text{Urban water supply conditions}\right] = f([\text{Total water resource modulus}]) \qquad (7.9)$$

2. **Evaluation steps**

The evaluation process is shown in Fig. 7.5.

Table 7.8 [GC process table] of land resource evaluation index

Step	Operation instruction	Input	Operation	Output	Description
1.	Slope calculation	[Height]	[**Description**] Slope calculation [**Keywords**] **KX_Slope**	[Slope 1] **DX2_PD1**	
2.	Reclassification	[Slope 1]	[**Description**] Raster reclassification (numerical range) [**Keywords**] N1: <V1llN2: V1 – V2l...lNn: ≥Vn **KX_Reclass(5: <2l 4: 2–6l 3: 6–15 l2: 15–25 l 1: ≥25)**	[Agricultural slope] DX2_NYPD	1(Very steep)2(Steep) 3(Common) 4(Relatively flat) 5(Flat)
3.	Modification operation	[Soil texture] [Agricultural slope]	[**Description**] Raster calculation If [Soil texture] ≥ 80%, the [Farming conditions 1] = 1; If 60% ≤ [Soil texture] < 80%, then [Farming conditions 1] = [Agricultural slope] – 1; [**Keywords**] Algebraic or logical expression **KX_RasCalculator(([R1] ≥ 80), 1% ([R1] ≥ 60) and ([R1] < 80), [R2] – 1%[R2])**	[Farming conditions 1] **DX2_NYGZTJ1**	
4.	Value modification	[Farming conditions 1]	[**Description**] Raster calculation If [Farming conditions 2] < 1, then [Farming conditions 2] = 1. [**Keywords**] Algebraic or logical expression **KX_RasCalculator(([R1] < 1), 1% [R1])**	[Farming conditions] **DX2_NYGZTJ**	5(Higher) 4(High) 3(Medium) 2(Low) 1(Lower)

(continued)

Table 7.8 (continued)

Step	Operation instruction	Input	Operation	Output	Description
5.	Reclassification	[Slope 1]	[**Description**] Raster reclassification (numerical range) [**Keywords**] N1: <V1|N2: V1 – V2|...|Nn: ≥Vn **KX_Reclass (5: <3| 4: 3–8| 3: 8–15 |2: 15–25 | 1: ≥25)**	[Town slope] **DX2_CZPD**	1(Very steep) 2(Steep) 3(Common) 4(Relatively flat) 5(Flat)
6.	Modification operation	[Height] [Town slope]	[**Description**] Raster calculation If [Height] ≥ 5000, then [Urban construction conditions 1] = 1; If 3500 < [Height] < 5000, then [Urban construction conditions 1] = [Town slope] − 1. [**Keywords**] Algebraic or logical expression **KX_RasCalculator (([R1] ≥ 5000), 1% ([R1] > 3500) and ([R1] < 5000), [R2] − 1%[R2])**	[Urban construction conditions 1] **DX2_CZJSTJ1**	
7.	Value modification	[Urban construction conditions 1]	[**Description**] Raster calculation [**Keywords**] Algebraic or logical expression **KX_RasCalculator(([R1] < 1), 1% [R1])**	[Urban construction conditions 2] **DX2_CZJSTJ2**	
8.	Undulation calculation	[Height]	[**Description**] Undulation calculation [**Keywords**] Neighborhood size **KX_Undulation(9)**	[Undulation] DX2_QFD	50 m * 50 as 9 30 m * 30 as 15 20 m * 20 as 21

(continued)

Table 7.8 (continued)

Step	Operation instruction	Input	Operation	Output	Description
9.	Modification operation	[Undulation] [Urban construction conditions 2]	**[Description]** Raster calculation If [Undulation] > 200, [Urban construction conditions 3] = [Urban construction conditions 2] − 2; If 100 < [Undulation] < 200, [Urban construction conditions 3] = [Urban construction conditions 2] − 1. **[Keywords]** Algebraic or logical expression **KX_RasCalculator (([R1] > 200), [R2] − 2 % ([R1] > 100) and ([R1] < 200) , [R2] − 1%[R2])**	[Urban construction conditions 3] **DX2_CZJSTJ3**	
10.	Value modification	[Urban construction conditions 3]	**[Description]** Raster calculation If [Urban construction conditions 3] < 1, [Urban construction conditions 3] = 1. **[Keywords]** Algebraic or logical expression **KX_RasCalculator(([R1] < 1), 1% [R1])**	[Urban construction conditions] **DX2_CZJSTJ**	5(Higher) 4(High) 3(Medium) 2(Low) 1(Lower)
11.	Make thematic map	[Farming conditions]	**[Description]** Make thematic map **[Keywords]** Replace list | Background list |Resolution | {Template} | {Drawing range} **KX_Mapping(C2|Boundary |200)**	[Thematic map of farming conditions] **DX2_NYGZTJ.emf**	

(continued)

Table 7.8   (continued)

Step	Operation instruction	Input	Operation	Output	Description
12.	Insert thematic map	[Thematic map of farming conditions]	**[Description]** Insert thematic map **[Keywords]** Picture heightl {Delete content} **KX_InsertPic (12)**	1	
13.	Produce statistical table	[Statistical layer] [Farming conditions]	**[Description]** Statistical table **[Keywords]** Display field # {Unit parameter} l Counted type list l {Data filtering mode} **KX_Statistic(Namel5,4,3,2,1)**	2	
14.	Make thematic map	[Urban construction conditions]	**[Description]** Make thematic map **[Keywords]** Replace list l Background list lResolution l {Template} l {Drawing range} **KX_Mapping(C3lBoundary l200)**	[Thematic map of urban construction conditions] **DX2_CZJSTJ.emf**	
15.	Insert thematic map	[Thematic map of urban construction conditions]	**[Description]** Insert thematic map **[Keywords]** Picture heightl {Delete content} **KX_InsertPic (13)**	3	
16.	Produce statistical table	[Statistical layer] [Urban construction conditions]	**[Description]** Statistical table **[Keywords]** Display field# {Unit parameter} l Counted type list l {Data filtering mode} **KX_Statistic (Namel5,4,3,2,1)**	4	

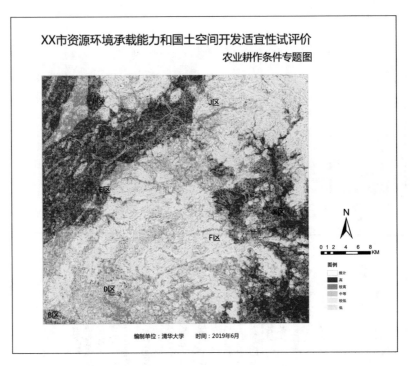

Fig. 7.3 The first example of thematic map

(1) Interpolate or rasterize (interpolation method is adopted when there are multiple meteorological station data, and rasterization method is adopted in other cases) [Precipitation] to obtain [Agricultural water supply conditions 1].

(2) [Agricultural water supply conditions 1] is reclassified according to Table 7.11 to obtain [Agricultural water supply conditions].

(3) Interpolate or rasterize (interpolation method is adopted when there are multiple data, and rasterization method is adopted in other cases) [Total Water resource modulus] to obtain [Urban water supply conditions 1].

(4) [Urban water supply conditions 1] is reclassified according to Table 7.12 to obtain [Urban water supply conditions].

7.3.2 Input and Control

Table 7.13 is the [Input and control table] for water resource evaluation.

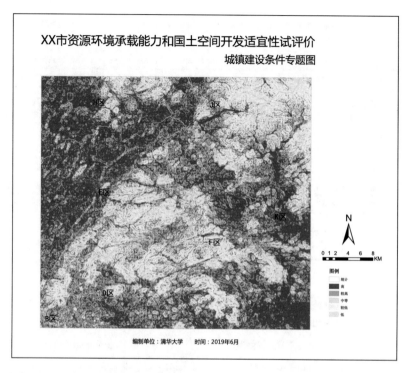

Fig. 7.4 The second example of thematic map

Table 7.9 The statistical table of agricultural cultivation conditions (unit: area, km²; proportion, %)

Region		Higher		High		Medium		Low		Lower	
		Area	Ratio	Area	Ratio	Area	Ratio	Area	Ratio	Area	Ratio
**City	B	1.02	8.1	4.70	37.4	5.16	41.0	1.56	12.4	0.14	1.1
	D	0.87	0.8	24.63	22.7	53.59	49.5	23.78	21.9	5.47	5.0
	E	93.29	20.5	121.73	26.8	106.92	23.5	91.38	20.1	41.48	9.1
	F	21.67	5.8	79.64	21.2	127.91	34.0	102.22	27.2	44.78	11.9
	H	51.19	27.4	67.79	36.3	40.68	21.8	20.25	10.8	6.81	3.6
	J	13.91	9.1	28.83	18.9	46.01	30.2	44.28	29.0	19.34	12.7
	K	109.10	26.2	116.17	27.8	78.32	18.8	74.23	17.8	39.45	9.5
	Total	291.04	17.0	443.48	26.0	458.60	26.8	357.72	20.9	157.46	9.2

Table 7.10 The statistical table of urban development conditions (unit: area, km²; proportion, %)

Region		Higher		High		Medium		Low		Lower	
		Area	Ratio	Area	Ratio	Area	Ratio	Area	Ratio	Area	Ratio
	B	3.34	26.5	6.36	50.6	2.41	19.2	0.43	3.4	0.04	0.3
	D	37.85	34.9	44.50	41.1	16.41	15.1	7.12	6.6	2.47	2.3
*	E	186.87	41.1	139.39	30.6	74.83	16.5	40.81	9.0	12.90	2.8
*	F	116.61	31.0	130.74	34.8	75.63	20.1	40.62	10.8	12.62	3.4
City	H	107.54	57.5	50.83	27.2	18.07	9.7	8.06	4.3	2.23	1.2
	J	36.83	24.1	43.55	28.5	35.88	23.5	25.67	16.8	10.44	6.8
	K	165.45	39.7	120.19	28.8	74.02	17.7	42.40	10.2	15.21	3.6
	Total	654.48	38.3	535.56	31.3	297.25	17.4	165.11	9.7	55.90	3.3

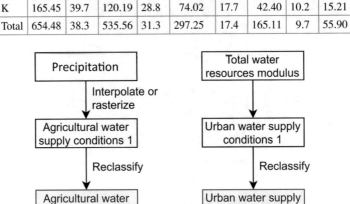

Fig. 7.5 GC process of water resource evaluation

Table 7.11 Precipitation grading reference thresholds

Precipitation classification	Arid	Semi-arid	Semi humid	Humid	Very humid
Precipitation	≤200	200–400	400–800	800–1200	>1200

Table 7.12 Graded reference thresholds of total water resource modulus (unit 10^4 m³/km²)

Classification of total water resource modulus	Bad	Poor	Common	Good	Excellent
Total water resource modulus	≤3	3–8	8–13	13–25	>25

Table 7.13 Water resource evaluation [Input and control table]

Serial No.	Layer name	Physical layer	Reference page	Value and description
1.	[Water resource]	DX3_SZY.shp		JYL-Precipitation, MS-Total water resource modulus
			[GC control]	

7.3.3   GC Process

Table 7.14 is the [GC process table] of water resource evaluation indexes.

Table 7.14   [GC process table] of water resource evaluation indexes

Step	Operation instruction	Input	Operation	Output	Description
1.	Rasterization	[Water resource]	**[Description]** Rasterization **[Keywords]** Operation field or value| {Grid size} **KX_FeatureToRaster (JYL)**	[Agricultural water supply conditions 1] DX3_NYGSTJ1	
2.	Reclassification	[Agricultural water supply conditions 1]	**[Description]** Raster reclassification (numerical range) **[Keywords]** N1: <V1|N2: V1 − V2|…|Nn: ≥Vn **KX_Reclass(1: <200| 2: 200–400| 3: 400–800 |4: 800–1200 | 5: ≥1200)**	[Agricultural water supply conditions 2] DX3_NYGSTJ2	
3.	Clip	[Agricultural water supply conditions 2] [Range layer]	**[Description]** Clip **[Keywords]** Clipped layer, Clip layer **KX_Clip**	[Agricultural water supply conditions] DX3_NYGSTJ	5(Very humid) 4(Humid) 3(Semi humid) 2(Semi-arid) 1(Drought)
4.	Rasterization	[Water resource]	**[Description]** Rasterization **[Keywords]** Operation field or value| {Grid size} **KX_FeatureToRaster(MS)**	[Urban water supply conditions 1] DX3_CZGSTJ	
5.	Reclassification	[Urban water supply conditions 1]	**[Description]** Raster reclassification (numerical range) **[Keywords]** N1: <V1|N2: V1 − V2|…|Nn: ≥Vn **KX_Reclass (1: <5| 2: 5–10| 3: 10–20 |4: 20–50 | 5: ≥50)**	[Urban water supply conditions 2] DX3_CZGSTJ2	
6.	Clip	[Urban water supply conditions 2] [Range layer]	**[Description]** Clip **[Keywords]** Clipped layer, Clip layer KX_Clip	[Urban water supply conditions] DX3_CZGSTJ	5(Excellent) 4(Good) 3(Common) 2(Poor) 1(Bad)
7.	Make thematic map	[Agricultural water supply conditions]	**[Description]** Make thematic map **[Keywords]** Replace list | Background list |Resolution | {Template} | {Drawing range} **KX_Mapping (C2|Boundary |200)**	[Thematic map of agricultural water supply conditions] DX3_NYGSTJ.emf	

(continued)

Table 7.14 (continued)

Step	Operation instruction	Input	Operation	Output	Description
8.	Insert thematic map	[Thematic map of agricultural water supply conditions]	**[Description]** Insert thematic map **[Keywords]** Picture heightl {Delete content} **KX_InsertPic (12)**	1	
9.	Produce statistical table	[Statistical layer] [Agricultural water supply conditions]	**[Description]** Statistical table **[Keywords]** Display field# {Unit parameter} l Counted type list l {Data filtering mode} **KX_Statistic (Namel5,4,3,2,1)**	2	
10.	Make thematic map	[Urban water supply conditions]	**[Description]** Make thematic map **[Keywords]** Replace list l Background list lResolution l {Template} l {Drawing range} **KX_Mapping (C3lBoundary l200)**	[Thematic map of urban water supply conditions] DX3_CZGSTJ.emf	
11.	Insert thematic map	[Thematic map of urban water supply conditions]	**[Description]** Insert thematic map **[Keywords]** Picture heightl {Delete content} **KX_InsertPic (12)**	3	
12.	Produce statistical table	[Statistical layer] [Urban water supply conditions]	**[Description]** Statistical table **[Keywords]** Display field# {Unit parameter} l Counted type list l {Data filtering mode} **KX_Statistic (Namel5,4,3,2,1)**	4	

7.4   Climate Evaluation

7.4.1   Evaluation Method

This model is referred to pages 32–33 of the *Technical Guidelines (June Edition)*.

1. **Evaluation methodology**

$$\begin{bmatrix} \text{Climatic conditions for agricultural production} \end{bmatrix} = f([\text{Photothermal condition}]) \tag{7.10}$$

$$[\text{Urban construction climate}] = f([\text{Comfort}]) \tag{7.11}$$

2. **Evaluation steps**

The evaluation process is shown in Fig. 7.6.

(1) The [Active accumulated temperature] is interpolated or rasterized to obtain [Photothermal condition 1].
(2) [Photothermal condition 1] is modified by [Elevation]. [Photothermal condition 2] is obtained by modifying the temperature decrease of 0.6°C for every 100 m height rise.
(3) [Photothermal condition 2] is reclassified according to Table 7.15 to obtain [Climatic conditions for agricultural production] (or [Photothermal condition]).
(4) According to the [12-month average temperature] of each meteorological station, the fuzzy mode is computed and spatialized to obtain the [Average temperature fuzzy mode].
(5) Interpolate or raster the [Average temperature fuzzy mode] (raster is adopted when there is only one station in the area), and modify it with [Elevation] to obtain the [Average temperature].
(6) According to the [12-month average humidity] of each meteorological station, the fuzzy mode is computed and spatialized to obtain the [Average humidity fuzzy mode].

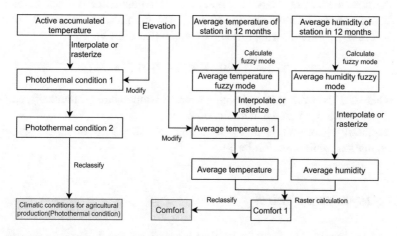

Fig. 7.6 GC flow chart of climate evaluation

Table 7.15 Active accumulated temperature rating reference thresholds

Classification of active accumulated temperature	Excellent	Good	Common	Poor	Bad
Active accumulated temperature	≥7600	5800–7600	4000–5800	1500–4000	<1500

Table 7.16 Comfort rating standards

Classification standards	Comfort levels
60–65	7(Very comfortable)
56–60 or 65–70	6
50–56 or 70–75	5
45–50 or 75–80	4
40–45 or 80–85	3
32–40 or 85–90	2
<32 or ≥90	1(Very uncomfortable)

(7) Interpolate or rasterize (rasterize is adopted when there is only one station in the area) the [Average humidity fuzzy mode] to obtain the [Average humidity].

(8) Using [Average temperature] and [Average humidity] to compute comfort according to the following formula to obtain [Comfort 1].

$$\text{THI} = \text{T} - 0.55 \times (1 - \text{f}) \times (\text{T} - 58) \tag{7.12}$$

where THI is comfort, T is monthly average temperature (Fahrenheit), and f is monthly average relative humidity of air (%).

(9) Reclassify [Comfort level 1] according to Table 7.16 to obtain the final evaluation result [Comfort level].

7.4.2   *Input and Control*

Table 7.17 shows the [Input and control table] of climate evaluation.

Table 7.17   Climate evaluation [Input and control table]

Serial No.	Layer name	Physical layer	Reference page	Value and description
1.	[Height]	DX \DX2\DX2_ GC		
2.	[Meteorology]	DX4_QXZD. shp		HDJW-Active accumulated temperature
		[GC control]		

7.4.3   GC Process

Table 7.18 shows the [GC Process table] of climate evaluation.

7.5   Environmental Evaluation

7.5.1   Evaluation Method

This model is referred to pages 34–39 of *Technical Guidelines (June Edition)*.

1. **Evaluation methodology**

$$[\text{Agricultural production environmental conditions}]$$
$$= f([\text{Soil environmental capacity}]) \tag{7.13}$$

$$[\text{Urban construction environmental conditions}] = \text{Min}([\text{Atmospheric}$$
$$\text{environmental capacity}], [\text{Water environmental capacity}]) \tag{7.14}$$

$$[\text{Atmospheric environmental capacity}] = \text{Min}([\text{Days of calm wind}],$$
$$[\text{Average wind speed}]) \tag{7.15}$$

2. **Evaluation steps**

The evaluation process is shown in Fig. 7.7.

(1)   Interpolate or rasterize (interpolation method is adopted when there are multiple station data, and rasterization method is adopted in other cases) [Soil pollution site] to obtain [Soil environmental capacity 1];

Table 7.18 [GC process table] of climate evaluation

Step	Operation instruction	Input	Operation	Output	Description
1.	Rasterization	[Meteorology]	[Description] Rasterization [Keywords] Operation field or value| {Grid size} KX_FeatureToRaster(HDJW)	[Photothermal conditions 1] DX4_GRTJ1	[Active accumulated temperature]
2.	Modification operation	[Photothermal conditions 1] [Height]	[Description] Raster calculation [Keywords] Algebraic or logical expression KX_RasCalculator([R1] − [R2]/100 * 0.6)	[Photothermal conditions 2] DX4_GRTJ2	
3.	Reclassification	[Photothermal conditions 2]	[Description] Reclassification (numerical range) [Keywords] N1: <V1|N2: V1 − V2|...|Nn: ≥Vn KX_Reclass (1: <1500| 2: 1500–4000| 3: 4000–5800 |4: 5800–7600 |5: ≥7600)	[Photothermal conditions 3] DX4_GRTJ3	
4.	Clip	[Photothermal conditions 3] [Range layer]	[Description] Clip [Keywords] Clipped layer, Clip layer KX_Clip	[Photothermal conditions] DX4_GRTJ	5(Excellent) 4(Good) 3(Common) 2(Poor) 1(Bad)
5.	Calculate the fuzzy mode from the 12-month temperature of meteorological stations	[meteorological phenomena]	[Description] Comfort calculation [Keywords] Name, comfort name, meteorological txt file (name, 12-month temperature list) KX_ZS (Name, WD,SSD1.txt)	[Comfort 1] DX4_SSD1.shp	

(continued)

Table 7.18 (continued)

Step	Operation instruction	Input	Operation	Output	Description
6.	Rasterization	[Comfort 1]	**[Description]** Rasterization **[Keywords]** Operation field or value {Grid size} **KX_FeatureToRaster (WD)**	[Comfort 2] DX4_SSD2	
7.	Modification operation	[Comfort 2] [Height]	**[Description]** Raster calculation **[Keywords]** Algebraic or logical expression **KX_RasCalculator([R1] – [R2]/100 * 0.6)**	[Comfort 3] DX4_SSD3	
8.	Convert to Fahrenheit	[Comfort 3]	**[Description]** Raster calculation **[Keywords]** Algebraic or logical expression **KX_RasCalculator([R1] * 9/5 + 32)**	[Comfort 31] DX4_SSD31 comfortable degree	Temperature (Fahrenheit)
9.	Calculate fuzzy mode from 12-month humidity of meteorological stations	[Meteorology]	**[Description]** Comfort calculation **[Keywords]** Name, Humidity name, Meteorological txt file (Name, 12-month relative humidity list) **KX_ZS (Name, SD, SSD2.txt)**	[Comfort 4] DX4_SSD4.shp	
10.	Rasterization	[Comfort 4]	**[Description]** Rasterization **[Keywords]** Operation field or value {Grid size} **KX_FeatureToRaster (SD)**	[Comfort 5] DX4_SSD5	humidity

(continued)

Table 7.18 (continued)

Step	Operation instruction	Input	Operation	Output	Description
11.	Comfort calculation	[Comfortable 31] [Comfort 5]	[Description] Raster calculation T − 0.55 × (1 − f) × (T − 58) [Keywords] Algebraic or logical expression **KX_RasCalculator([R1] − 0.55 * (1 − [R2]/100) * ([R1] − 58))**	[Comfort 6] DX4_SSD6	
12.	Reclassification	[Comfort 6]	[Description] Reclassification (numerical range) [Keywords] N1: <V1|N2: V1 − V2|...|Nn: ≥Vn **KX_Reclass(1: <32| 2: 32–40| 3: 40–45| 4: 45–50| 5: 50–56| 6: 56–60| 7: 60–65| 8: 65–70| 9: 70–75| 10: 75–80|11: 80–85| 12: 85–90 | 13: ≥90)**	[Comfort 7] DX4_SSD7	
13.	Reclassification	[Comfort 7]	[Description] Reclassification (class modification) [Keywords] N1: C1, C2, ...|N2: C3, C4, ...|... **KX_Reclass(1: 1, 13|2: 2, 12|3: 3, 11|4, 10|5: 5, 9|6: 6, 8|7: 7)**	[Comfort 8] DX4_SSD8	
14.	Clip	[Comfort 8] [Range layer]	[Description] Clip [Keywords] Clipped layer, Clip layer **KX_Clip**	[Comfort] DX4_SSD	7(Very comfortable)/6/5/4/3/2/1 (Very uncomfortable)

(continued)

Table 7.18   (continued)

Step	Operation instruction	Input	Operation	Output	Description
15.	Make thematic map	[Photothermal conditions]	**[Description]** Make thematic map **[Keywords]** Replace list I Background list IResolution I {Template} I {Drawing range} **KX_Mapping(C2IBoundary I200)**	[Classification diagram of Photothermal conditions] DX4_GRTJ.emf	
16.	Insert thematic map	[Classification diagram of photothermal conditions]	**[Description]** Insert thematic map **[Keywords]** Picture heightl {Delete content} **KX_InsertPic (12)**	1	
17.	Produce statistical table	[Statistical layer] [Photothermal conditions]	**[Description]** Make statistical table **[Keywords]** Display field# {Unit parameter} I Counted type list I {Data filtering mode} **KX_Statistic (NameI5,4,3,2,1)**	2	
18.	Make thematic map	[Comfort]	**[Description]** Make thematic map **[Keywords]** Replace list I Background list IResolution I {Template} I {Drawing range} **KX_Mapping(SSDIBoundary I200)**	[Comfort index grading chart] DX4_SSD.emf	
19.	Insert thematic map	[Comfort index grading chart]	**[Description]** Insert thematic map **[Keywords]** Picture heightl {Delete content} **KX_InsertPic (13)**	3	
20.	Produce statistical table	[Statistical layer] [Comfort]	**[Description]** Make statistical table **[Keywords]** Display field# {Unit parameter} I Counted type list I {Data filtering mode} **KX_Statistic (NameI7,6,5,4,3,2,1)**	4	

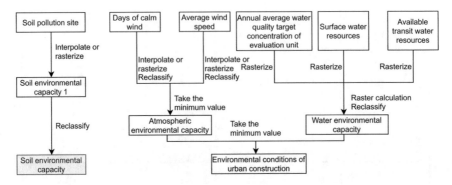

Fig. 7.7 GC process of environmental evaluation

(2) [Soil environmental capacity 1] was reclassified to obtain [Agricultural production environmental conditions];

(3) Interpolate or rasterize (interpolation method is adopted when there are multiple
station data, and rasterization method is adopted in other cases) [Days of
calm wind] and [Average wind speed] to obtain [Atmospheric environmental
capacity];

(4) The water environment capacity is obtained by rasterizing [Annual average
water quality target concentration of the evaluation unit], [surface water
resources] and [Available transit water resources] respectively, and their
respective results are computed according to formula (7.16);

$$[\text{Water environmental capacity}] = [\text{Annual average water quality target}$$
$$\text{concentration of the evaluation unit}] \times ([\text{Surface water resources}]$$
$$+[\text{Available transit water resources}]) \tag{7.16}$$

(5) Take the minimum values of [Atmospheric environmental capacity] and
[Water environmental capacity] to obtain [Environmental conditions of urban
construction].

7.5.2 Input and Control

Table 7.19 is the [Input and control table] of environmental evaluation.

7.5.3 GC Process

Table 7.20 shows the [GC Process table] of environmental evaluation indexes.

Table 7.19 Environmental evaluation [Input and control table]

Serial No.	Layer name	Physical layer	Reference page	Value and description
1.	[Soil pollution site]	DX5_TRWR.shp	34	TRWR-Soil pollution
2.	[Weather station]	DX5_QXZD.shp	34	JFRS-Calm days , PJFS-Average wind speed
3.	[Annual average water quality target concentration of evaluation unit]	DX5_MBLD.shp	38	MBLD-Target concentration
4.	[Surface water resources]	DX5_DBS.shp	38	DBS-Surface water resources
5.	[Available transit water resources]	DX5_GJS.shp	38	KLYGJS-Available transit water resources
		[GC control]		

7.6 Disaster Evaluation

7.6.1 Evaluation Method

This model is referred to pages 40–45 of *Technical Guidelines (June Edition)*.

1. **Evaluation methodology**

$$[\text{Meteorological disaster risk}] = \text{Max}([\text{Risk of drought disaster}],$$
$$[\text{Risk of flood disaster}], [\text{Risk of low temperature and cold disaster}]) \quad (7.17)$$

$$[\text{Disaster risk}] = \text{Max}([\text{Seismic risk}], [\text{Geological hazard susceptibility}],$$
$$[\text{Risk of storm surge disaster}]) \quad (7.18)$$

$$[\text{Seismic risk}] = \text{Max}([\text{Active fault}], [\text{Peak ground motion acceleration}]) \quad (7.19)$$

$$[\text{Geological hazard susceptibility}] = \text{Max}([\text{Susceptibility of landslide}$$
$$\text{and debris flow}], [\text{Susceptibility of land subsidence}],$$
$$[\text{Susceptibility of ground collapse}]) \quad (7.20)$$

2. **Evaluation steps**

The evaluation process is shown in Fig. 7.8.

Table 7.20 [GC process table] of environmental evaluation indexes

Steps	Operation instruction	Input	Operation	Output	Description
1.	Spatial interpolation	[Soil pollution site]	**[Description]** Spatial interpolation [**Keywords**] Interpolation field | {Interpolation method} | {Number of search points} **KX_Interpolation(TRWR)**	[Soil environmental capacity 1] DX5_TRHJRL1	
2.	Reclassification	[Soil environmental capacity 1]	**[Description]** Reclassification (numerical range) [**Keywords**] N1: <V1|N2: V1 – V2|...|Nn: ≥Vn **KX_Reclass(3: <100|2: 100–150 | 1: ≥150)**	[Soil environmental capacity 2] DX5_TRHJRL2	3(Higher) 2(Medium) 1(Lower)
3.	Clip	[Soil environmental capacity 2] [Range layer]	**[Description]** Clip [**Keywords**] **KX_Clip**	[Soil environmental capacity] DX5_TRHJRL	
4.	Spatial interpolation	[Weather station]	**[Description]** Spatial interpolation [**Keywords**] Interpolation field| {Interpolation method} | {Number of search points} **KX_Interpolation (JFRS)**	[Calm days 1] Calm days DX5_JFRS1	
5.	Reclassification	[Calm days 1]	**[Description]** Reclassification (numerical range) [**Keywords**] N1: <V1|N2: V1 – V2|...|Nn: ≥Vn KX_Reclass(1: <5| 2: 5–10| 3: 10–20 |4: 20–30 | 5: ≥30)	[Calm days] Calm days DX5_JFRS	5(Higher) 4(High) 3(Medium) 2(Low) 1(Lower)

(continued)

Table 7.20 (continued)

Steps	Operation instruction	Input	Operation	Output	Description
6.	Spatial interpolation	[Weather station]	[Description] Spatial interpolating [Keywords] Interpolation field| {Interpolation method} | {Number of search points} **KX_Interpolation (PJFS)**	[Average wind speed 1] DX5_PJFS1	
7.	Reclassification	[Average wind speed 1]	[Description] Reclassification (numerical range) [Keywords] N1: <V1|N2: V1 – V2|…|Nn: ≥Vn **KX_Reclass(1: <1| 2: 1–2| 3: 2–3 |4: 3–5 | 5: ≥5)**	[Average wind speed] DX5_PJFS	5(Higher) 4(High) 3(Medium) 2(Low) 1(Lower)
8.	Take the minimum value	[Calm days] [Average wind speed]	[Description] Take the minimum value [Keywords] **KX_Min**	[Atmospheric environmental capacity 1] DX5_DQHJRL1	
9.	Clip	[Atmospheric environmental capacity 1] [Range layer]	[Description] Clip [Keywords] Clipped layer, Clip layer **KX_Clip**	[Atmospheric environmental capacity] DX5_DQHJRL	5(Higher) 4(High) 3(Medium) 2(Low) 1(Lower)
10.	Rasterization	[Annual average water quality target concentration of evaluation unit]	[Description] Rasterization [Keywords] Operation field or value| {Grid size} **KX_FeatureToRaster(MBLD)**	[Annual average water quality target concentration of evaluation unit R] DX5_MBLD	

(continued)

Table 7.20 (continued)

Steps	Operation instruction	Input	Operation	Output	Description
11.	Rasterization	[Surface water resources]	[Description] Rasterization [Keywords] Operation field or value{Grid size} **KX_FeatureToRaster (DBS)**	[Surface water resources R] DX5_DBS	
12.	Rasterization	[Available transit water resources]	[Description] Rasterization [Keywords] Operation field or value{Grid size} **KX_FeatureToRaster(KLYGJS)**	[Available transit water resources R] DX5_GJS	
13.	Raster calculation	[Annual average water quality target concentration of evaluation unit R] [Surface water resources R] [Available transit water resources R]	[Description] Raster calculation [Keywords] Algebraic or logical expression **KX_RasCalculator ([R1] * [R2] + [R3])**	[Water environmental capacity 1] DX5_SHJRL1	
14.	Reclassification	[Water environmental capacity 1]	[Description] Reclassification (equal distance) [Keywords] N1lN2l...lNn **KX_Reclass(5l4l3l2l1)**	[Water environmental capacity 2] DX5_SHJRL2	
15.	Clip	[Water environmental capacity 2] [Range layer]	[Description] Clip [Keywords] Clipped layer, Clip layer **KX_Clip**	[Water environmental capacity] DX5_SHJRL	5(Higher) 4(High) 3(Medium) 2(Low) 1(Lower)
16.	Take the minimum value	[Atmospheric environmental capacity] [Water environmental capacity]	[Description] Take the minimum value [Keywords] **KX_Min**	[Environmental conditions of urban construction1] DX5_CZHJTJ1	

(continued)

Table 7.20 (continued)

Steps	Operation instruction	Input	Operation	Output	Description
17.	Clip	[Environmental conditions of urban construction 1] [Range layer]	**[Description]** Clip **[Keywords]** Clipped layer, Clip layer **KX_Clip**	[Environmental conditions of urban construction] DX5_CZHJTJ	5(Excellent) 4(Good) 3(Medium) 2(Poor) 1(Bad)
18.	Make thematic map	[Soil environmental capacity]	**[Description]** Make thematic map **[Keywords]** Replace list ǀ Background list ǀResolution ǀ {Template} ǀ {Drawing range} **KX_Mapping (C1ǀBoundary ǀ200)**	[Thematic map of soil environmental capacity] DX5_TRHJRL.emf	
19.	Insert thematic map	[Thematic map of soil environmental capacity]	**[Description]** Insert thematic map **[Keywords]** Picture heightǀ {Delete content} **KX_InsertPic (12)**	1	
20.	Make statistical table	[Statistical layer] [Soil environmental capacity]	**[Description]** Make statistical table **[Keywords]** Display field# {Unit parameter} ǀ Counted type list ǀ {Data filtering mode} **KX_Statistic (Nameǀ5,4,3,2,1)**	2	
21.	Make thematic map	[Atmospheric environmental capacity]	**[Description]** Make thematic map **[Keywords]** Replace list ǀ Background list ǀResolution ǀ {Template} ǀ {Drawing range} **KX_Mapping (C2ǀBoundary ǀ200)**	[Thematic map of atmospheric environmental capacity] DX5_DQHJRL.emf	

(continued)

Table 7.20 (continued)

Steps	Operation instruction	Input	Operation	Output	Description
22.	Insert thematic map	[Thematic map of atmospheric environmental capacity]	[Description] Insert thematic map [Keywords] Picture heightl {Delete content} **KX_InsertPic (12)**	3	
23.	Make statistical table	[Statistical layer] [Atmospheric environmental capacity]	[Description] Make statistical table [Keywords] Display field# {Unit parameter} l Counted type list l {Data filtering mode} **KX_Statistic(Namel5,4,3,2,1)**	4	
24.	Make thematic map	[Water environmental capacity]	[Description] Make thematic map [Keywords] Replace list l Background list lResolution l {Template} l {Drawing range} **KX_Mapping(C3lBoundary l200)**	[Thematic map of water environmental capacity] DX5_SHJRL.emf	
25.	Insert thematic map	[Thematic map of water environmental capacity]	[Description] Insert thematic map [Keywords] Picture heightl {Delete content} **KX_InsertPic (12)**	5	
26.	Make statistical table	[Statistical layer] [Water environmental capacity]	[Description] Make statistical table [Keywords] Display field# {Unit parameter} l Counted type list l {Data filtering mode} **KX_Statistic (Namel5,4,3,2,1)**	6	

(continued)

Table 7.20 (continued)

Steps	Operation instruction	Input	Operation	Output	Description
27.	Make#	[Environmental conditions of urban construction]	[Description] Make thematic map [Keywords] Replace list l Background list lResolution l {Template} l {Drawing range} **KX_Mapping (C2lBoundary l200)**	[Thematic map of environmental conditions for urban construction] DX5_CZHJTJ .emf	
28.	Insert thematic map	[Thematic map of environmental conditions for urban construction]	[Description] Insert thematic map [Keywords] Picture heightl {Delete content} **KX_InsertPic (12)**	7	
29.	Make statistical table	[Statistical layer] [Environmental conditions of urban construction]	[Description] Make statistical table [Keywords] Display field# {Unit parameter} l Counted type list l {Data filtering mode} **KX_Statistic (Namel5,4,3,2,1)**	8	

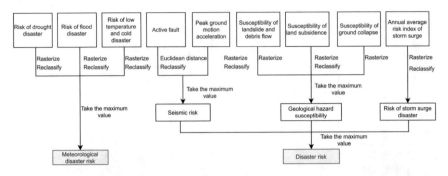

Fig. 7.8 GC flow chart of disaster evaluation

(1) Rasterize [Risk of drought disaster], [Risk of flood disaster] and [Risk of low temperature and cold disaster] respectively, and classify them according to their respective classification standards, and obtain [Meteorological disaster risk 1], [Meteorological disaster risk 2] and [Meteorological disaster risk 3].

(2) Take the maximum values of [Meteorological disaster risk 1], [Meteorological disaster risk 2] and [Meteorological disaster risk 3] to obtain the [Meteorological disaster risk].

(3) [Active fault] is reclassified according to Table 7.21 after Euclidean distance operation to obtain [Seismic risk 1] (Table 7.21).

(4) After [Peak ground motion acceleration] is rasterized, it is reclassified according to Table 7.22 to obtain [Seismic risk 2] (Table 7.22).

(5) Taking the maximum values of [Seismic risk 1] and [Seismic risk 2] to obtain the [Seismic risk];

(6) [Geological hazard susceptibility 1] and [Geological hazard susceptibility 2] can be obtained after [Susceptibility of landslide and debris flow] and [Susceptibility of ground subsidence] are rasterized.

(7) After [Susceptibility of land subsidence] is rasterized, it is reclassified according to Table 7.23 to obtain [Geological hazard susceptibility 3] (Table 7.23).

(8) Taking the maximum values of [Geological hazard susceptibility 1], [Geological hazard susceptibility 2] and [Geological hazard susceptibility 3], the [Geological hazard susceptibility] is obtained.

Table 7.21 Classification table of safe distance of active faults or ground fissures (unit: m)

Risk level	Higher	High	Medium	Lower
Distance from fracture	<100	100–200	200–400	≥400

Table 7.22 Grading table of ground motion acceleration (unit: g)

Risk level	Not easy	Low	Medium	High
Peak acceleration of ground motion	<0.05	0.05–0.2	0.2–0.4	≥0.4

Table 7.23   Surface subsidence classification table (unit: mm)

Grade	Not easy	Low	Medium	High
Accumulated settlement	<200	200–800	800–1600	≥1800

Table 7.24   Grading reference thresholds of annual average risk index of storm surge disaster

Risk level	Lower	Low	High	Higher
Annual average risk index of storm surge	<2.0	2.0–3.5	3.5–7.0	≥7.0

(9)   After the [Annual average risk index of storm] is rasterized, it is reclassified according to Table 7.24 to obtain the [Risk of storm surge disaster].

(10)   Take the maximum values of [Seismic risk], [Geological hazard susceptibility] and [Risk of storm surge disaster] to obtain [Disaster risk].

7.6.2   Input and Control Table

Table 7.25 shows the [Input and control table] of disaster evaluation.

Table 7.25   Disaster evaluation [Input and control table]

Serial No.	Layer name	Physical layer	Reference page	Value and description
1.	[Risk of drought disaster]	DX6_GHPL.shp		GHPL-Risk of drought disaster
2.	[Risk of flood disaster]	DX6_HLPL.shp		HLPL-Risk of flood disaster
3.	[Risk of low temperature and cold disaster]	DX6_DWLPL.shp		DWLPL-Risk of low temperature and cold disaster
4.	[Active fault]	DX6_HDDLD.shp		Active fault
5.	[Peak ground motion acceleration]	DX6_DZJSD.shp		DZJSD (Peak ground motion acceleration)
6.	[Susceptibility of landslide and debris flow]	DX6_BTYFX.shp		BTYFX-1(Not easy)/2(Low)/3(Medium)/4(High)
7.	[Susceptibility of land subsidence]	DX6_CJYFX.shp		CJSL (Subsidence rate)
8.	[Susceptibility of ground collapse]	DX6_TXYFX.shp		TXYFX-1(Not easy)/2(Low)/3(Medium)/4(High)
9.	[Annual average risk index of storm surge]	DX6_FBCZS.shp		
		[GC control]		

7.6.3   GC Process

Table 7.26 shows the [Calculate process table] of disaster evaluation.

Table 7.26 [Calculate process table] for disaster evaluation

Step	Operation instruction	Input	Operation	Output	Description
1.	Rasterization	[Risk of drought disaster]	**[Description]** Rasterization **[Keywords]** Operation field or value| {Grid size} **KX_FeatureToRaster (GHPL)**	[Risk of drought disaster 1] DX6_GHWXX1	
2.	Reclassification	[Risk of drought disaster 1]	**[Description]** Reclassification (numerical range) **[Keywords]** N1: <V1|N2: V1 − V2|...|Nn: ≥Vn **KX_Reclass (1: <20|2: 20–40|3: 40–60|4: 60–80|5: ≥80)**	[Risk of drought disaster] DX6_GHWXX	5(Higher) 4(High) 3(Medium) 2(Low) 1(Lower)
3.	Rasterization	[Risk of flood disaster]	**[Description]** Rasterization **[Keywords]** Operation field or value| {Grid size} **KX_FeatureToRaster (HLPL)**	[Risk of flood disaster 1] DX6_HLWXX1	
4.	Reclassification	[Risk of flood disaster 1]	**[Description]** Reclassification (numerical range) **[Keywords]** N1: <V1|N2: V1 − V2|...|Nn: ≥Vn **KX_Reclass (1: <20|2: 20–40|3: 40–60|5: ≥80)**	[Risk of flood disaster] DX6_HLWXX	5(Higher) 4(High) 3(Medium) 2(Low) 1(Lower)
5.	Rasterization	[Risk of low temperature and cold disaster]	**[Description]** Rasterization **[Keywords]** Operation field or value| {Grid size} **KX_FeatureToRaster (DWLPL)**	[Risk of low temperature and cold disaster 1] DX6_DWLWXX1	
6.	Reclassification	[Risk of low temperature and cold disaster 1]	**[Description]** Reclassification (numerical range) **[Keywords]** N1: <V1|N2: V1 − V2|...|Nn: ≥Vn **KX_Reclass (1: <20|2: 20–40|3: 40–60|4: 60–80|5: ≥80)**	[Risk of low temperature and cold disaster] DX6_DWLWXX	5(Higher) 4(High) 3(Medium) 2(Low) 1(Lower)

(continued)

Table 7.26   (continued)

Step	Operation instruction	Input	Operation	Output	Description
7.	Take the maximum	[Risk of drought disaster] [Risk of flood disaster] [Risk of low temperature and cold disaster]	**[Description]** Take the maximum value **[Keywords]** **KX_Max**	[Meteorological disaster risk 1] DX6_QXZHFX1	
8.	Clip	[Meteorological disaster risk 1] [Range layer]	**[Description]** Clip **[Keywords]** Clipped layer, Clip layer **KX_Clip**	[Meteorological disaster risk] DX6_QXZHFX	5(Higher) 4(High) 3(Medium) 2(Low) 1(Lower)
9.	Euclidean distance + Reclassification	[Active fault]	**[Description]** Filter + Euclidean distance + Reclassification **[Keywords]** {Operation field | Filter list} % Reclassification information **KX_SelDisReClass (4: <100| 3: 100–200| 2: 200–400| 1: ≥400)**	[Seismic risk 1] DX6_DZWXX1	
10.	Rasterization	[Peak ground motion acceleration]	**[Description]** Rasterization **[Keywords]** Operation field or value| {Grid size} **KX_FeatureToRaster (DZJSD)**	[Seismic risk 21] DX6_DZWXX21	
11.	Reclassification	[Seismic risk 21]	**[Description]** Reclassification (numerical range) **[Keywords]** N1: <V1|N2: V1 − V2|…|Nn: ≥Vn **KX_Reclass (1: <0.05|2: 0.05–0.1|3: 0.1–0.15|4: 0.15–0.3|5: ≥0.3)**	[Seismic risk 2] DX6_DZWXX2	
12.	Take the maximum	[Seismic risk 1] [Seismic risk 2]	**[Description]** Take the maximum value **[Keywords]** **KX_Max**	[Seismic risk] DX6_DZWXX	5(Higher) 4(High) 3(Medium) 2(Low) 1(Lower)
13.	Rasterization	[Susceptibility of landslide and debris flow]	**[Description]** Rasterization **[Keywords]** Operation field or value | {Grid size} **KX_FeatureToRaster (BTYFX)**	[Geological hazard susceptibility 1] DX6_DZZHYFX1	

(continued)

Table 7.26 (continued)

Step	Operation instruction	Input	Operation	Output	Description
14.	Rasterization	[Susceptibility of land subsidence]	**[Description]** Rasterization **[Keywords]** Operation field or value | {Grid size} **KX_FeatureToRaster (CJSL)**	[Geological hazard susceptibility 21] DX6_DZZHYFX21	
15.	Reclassification	[Geological hazard susceptibility 21]	**[Description]** Reclassification (numerical range) **[Keywords]** N1: <V1|N2: V1 − V2|...|Nn: ≥Vn **KX_Reclass (1: <10|2: 10–30|3: 30–50|4: ≥50)**	[Geological hazard susceptibility 2] DX6_DZWXX2	
16.	Rasterization	[Susceptibility of Ground collapse]	**[Description]** Rasterization **[Keywords]** Operation field or value | {Grid size} **KX_FeatureToRaster (TXYFX)**	[Geological hazard susceptibility 3] DX6_DZZHYFX3	
17.	Take the maximum	[Geological hazard susceptibility 1] [Geological hazard susceptibility 2] [Geological hazard susceptibility 3]	**[Description]** Take the maximum value **[Keywords]** **KX_Max**	[Geological hazard susceptibility] DX6_DZZHYFX	4(Higher) 3(High) 2(Medium) 1(Lower)
18.	Rasterization	[Annual average risk index of storm surge]	**[Description]** Rasterization **[Keywords]** Operation field or value | {Grid size} **KX_FeatureToRaster (FBCZS)**	[Risk of storm surge disaster 1] DX6_FBCZHWXX1	
19.	Reclassification	[Risk of storm surge disaster 1]	**[Description]** Reclassification (numerical range) **[Keywords]** N1: <V1|N2: V1 − V2|...|Nn: ≥Vn **KX_Reclass (1: <2|2: 2–3.5|3: 3.5–7|4: ≥7)**	[Risk of storm surge disaster] DX6_FBCZHWXX	4(Higher) 3(High) 2(Medium) 1(Lower)
20.	Take the maximum	[Seismic risk] [Geological hazard susceptibility] [Risk of storm surge disaster]	**[Description]** Take the maximum value **[Keywords]** **KX_Max**	[Disaster risk 1] DX6_ZHWXX1	
21.	Clip	[Disaster risk 1] [Range layer]	**[Description]** Clip **[Keywords]** Clipped layer, Clip layer **KX_Clip**	[Disaster risk] DX6_ZHWXX	4(Higher) 3(High) 2(Medium) 1(Lower)

(continued)

Table 7.26 (continued)

Step	Operation instruction	Input	Operation	Output	Description
22.	Make thematic map	[Meteorological disaster risk]	**[Description]** Make thematic map **[Keywords]** Replace list | Background list |Resolution | {Template} | {Drawing range} **KX_Mapping (C2|Boundary |200)**	[Thematic map of meteorological disaster risk] DX6_QXZHFX.emf	
23.	Insert thematic map	[Thematic map of meteorological disaster risk]	**[Description]** Insert thematic map **[Keywords]** Picture height| {Delete content} **KX_InsertPic (12)**	1	
24.	Make statistical table	[Statistical layer] [Meteorological disaster risk]	**[Description]** Make statistical table **[Keywords]** Display field# {Unit parameter} | Counted type list | {Data filtering mode} **KX_Statistic (Name|5,4,3,2,1)**	2	
25.	Make thematic map	[Disaster risk]	**[Description]** Make thematic map **[Keywords]** Replace list | Background list |Resolution | {Template} | {Drawing range} **KX_Mapping (C2|Boundary |200)**	[Thematic map of disaster risk] DX6_ZHWXX.emf	
26.	Insert thematic map	[Thematic map of disaster risk]	**[Description]** Insert thematic map **[Keywords]** Picture height| {Delete content} **KX_InsertPic (12)**	3	
27.	Make statistical table	[Statistical layer] [Disaster risk]	**[Description]** Make statistical table **[Keywords]** Display field# {Unit parameter} | Counted type list | {Data filtering mode} **KX_Statistic (Name|4,3,2,1)**	4	

7.7 Location Evaluation (Provincial Level)

7.7.1 Evaluation Method

This model refers to page 46 in *Technical Guidelines (June Edition)*.

1. **Evaluation methodology**

$$[\text{Location advantage}] = f([\text{Traffic distance to central city}]) \qquad (7.21)$$

2. Evaluation steps

The evaluation process is shown in Fig. 7.9.

(1) Select Level 1 cities from [Central city], carry out Euclidean distance operation, and reclassify them according to Table 7.27 to obtain [Traffic distance classification of central city 1].
(2) Select level 2 cities from [Central city], carry out Euclidean distance operation, and reclassify them according to Table 7.27 to obtain [Traffic distance classification of central city 2].
(3) Select 3-level cities from [Central city], carry out Euclidean distance operation, and reclassify them according to Table 7.27 to obtain [Traffic distance classification of central city 3].
(4) Calculate the sum of [Traffic distance classification of central city 1], [Traffic distance classification of central city 2] and [Traffic distance classification of central city 3], and reclassify it to obtain the [Location advantage].

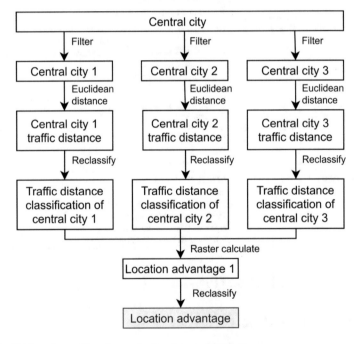

Fig. 7.9 GC flow chart of location evaluation (provincial level)

Table 7.27 Classification table of traffic distance from central city (unit: km)

Evaluation grade	Excellent	Good	Common	Poor	Bad
Distance from level 1 central city	<80	80–160	160–280	280–400	≥400
Distance from level 2 central city	<40	40–120	120–200	200–320	≥320
Distance from level 3 central city	<40	40–120	120–160	160–240	≥240

Table 7.28 [Input and control table] of comprehensive advantage evaluation (provincial level)

Serial No.	Layer name	Physical layer	Reference page	Value and description
1.	[Central city]	DX71_ZXCS.shp	46	DJ-Central city level
		[GC control]		

7.7.2 Input and Control

Table 7.28 shows the [Input and control table] of comprehensive advantage evaluation (provincial level).

7.7.3 GC Process

Table 7.29 is the [GC Process table] of comprehensive advantage evaluation (provincial level).

7.8 Location Evaluation (City and County)

7.8.1 Evaluation Method

This model is referred to pages 46–51 in *Technical Guidelines (June Edition)*.

1. **Evaluation methodology**

$$[\text{Comprehensive advantage}] = f([\text{Location conditions}], \\ [\text{Traffic network density}]) \qquad (7.22)$$

$$[\text{Location conditions}] = f([\text{Accessibility of traffic trunk lines}], \\ [\text{Accessibility of central urban area}], [\text{Accessibility of traffic hub}], \\ [\text{Accessibility of surrounding central cities}]) \qquad (7.23)$$

2. **Evaluation steps**

The evaluation process is shown in Fig. 7.10.

(1) Using [Road] to obtain [Accessibility of first-class highway], [Accessibility of secondary highway], [Accessibility of class III highway] and [Accessibility of class IV highway] respectively through filtering, Euclidean distance, reclassification, and other operations.

(2) Sum up the above computed results and reclassify them into 5 levels to obtain [Accessibility of traffic trunk lines].

Table 7.29 [GC process table] of comprehensive advantage evaluation (provincial level)

Step	Operation instruction	Input	Operation	Output	Description
1.	Filter + Euclidean distance + Reclassification	[Central city]	**[Description]** Filter + Euclidean distance + Reclassification **[Keywords]** {Operation field	Filter list} % Reclassification information **KX_SelDisReClass (DJl1 % 5:** <80,000| 4: 80,000–160,000| 3: 160,000–280,000| 2: 280,000–400,00011: ≥400,000)	[Central city 1] DX71_ZXCS1	Calculated by 80 km/h 1|2|3.5|5
2.	Filter + Euclidean distance + Reclassification	[Central city]	**[Description]** Filter + Euclidean distance + Reclassification **[Keywords]** {Operation field	Filter list} % Reclassification information **KX_SelDisReClass (DJl1 % 5:** <40,000| 4: 40,000–120,000| 3: 120,000–200,000| 2: 200,000–320,00011: ≥320,000)	[Central city 2] DX71_ZXCS2	Calculated by 80 km/h 0.5|1.5|2.5|4
3.	Filter + Euclidean distance + Reclassification	[Central city]	**[Description]** Filter + Euclidean distance + Reclassification **[Keywords]** {Operation field	Filter list} % Reclassification information **KX_SelDisReClass (DJl1 % 5:** <40,000| 4: 40,000–120,000| 3: 120,000–160,000| 2: 160,000–240,00011: ≥240,000)	[Central city 3] DX71_ZXCS3	Calculated by 80 km/h 0.5|1.5|2.0|3

(continued)

Table 7.29 (continued)

Step	Operation instruction	Input	Operation	Output	Description
4.	Raster calculation	[Central city 1] [Central city 2] [Central city 3]	[Description] Raster calculation [Keywords] Algebraic or logical expression KX_RasCalculator ([R1] + [R2] + [R3])	[Location advantage 1] DX71_QWYSD1	
5.	Reclassification	[Location advantage 1]	[Description] Reclassification (equal distance), [Keywords] N1|N2|...|Nn KX_Reclass(1|2|3|4|5)	[Location advantage 2] DX71_QWYSD2	
6.	Clip	[Location advantage 2] [Range layer]	[Description] Clip [Keywords] Clipped layer, Clip layer KX_Clip	[Location advantage] DX71_QWYSD	
7.	Make thematic map	[Location advantage]	[Description] Make thematic map [Keywords] Replace list	Background list	Resolution	{Template}	{Drawing range} KX_Mapping(C2|Boundary|200)	[Thematic map of location advantage] DX71_QWTJ.emf	
8.	Insert thematic map	[Thematic map of location advantage]	[Description] Insert thematic map [Keywords] Picture height| {Delete content} KX_InsertPic (12)	1	
9.	Make statistical table	[Statistical layer] [Location advantage]	[Description] Make statistical table [Keywords] Display field# {Unit parameter}	Counted type list	{Data filtering mode} KX_Statistic (Name	5,4,3,2,1)	2	

(3) Using [Road network] to compute the service scope of the [Central city], the [Accessibility of central urban area] is obtained.

(4) Using [Road network] to compute the service scope of [Airport], [Railway station], [Port], [Highway hub], [Freeway entrance] and [Urban rail transit station] respectively.

(5) The results computed above are summed up and reclassified into 5 levels to obtain [Accessibility of transportation hub].

(6) The [Accessibility of traffic trunk lines], [Accessibility of central urban area] and [Accessibility of traffic hub] are summed up and reclassified to obtain [Location conditions].

(7) Calculate the [Road] through the kernel density and reclassify it to obtain the [Traffic network density].

(8) Calculate [Location conditions] and [Traffic network density] according to Table 7.30 to obtain the evaluation result of [Comprehensive advantage].

7.8.2   Input and Control Table

Table 7.31 shows the [Input and control table] of comprehensive Advantage evaluation (city and county).

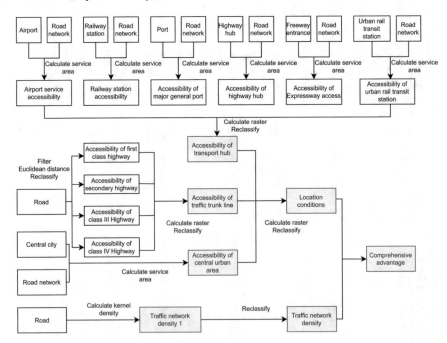

Fig. 7.10 GC flow chart for comprehensive advantage

Table 7.30   Reference discriminant matrix of location advantage degree

Traffic network density	Location condition				
	Excellent	Good	Common	Poor	Bad
Higher	Higher	Higher	High	Medium	Lower
High	Higher	Higher	High	Low	Lower
Common	Higher	High	Medium	Low	Lower
Low	High	High	Medium	Lower	Lower
Lower	Medium	Medium	Low	Lower	Lower

Table 7.31   [Input and control table] of comprehensive advantage evaluation (city and county)

Serial No.	Layer name	Physical layer	Reference page	Value and description
1.	[Road]	DX72_ROAD.shp		Line layer, DJ (1,2,3,4)
2.	[Central city]	DX72_ZXCQ.shp		Point layer
3.	[Airport]	DX72_C1A.shp		Point layer
4.	[Railway station]	DX72_C2R.shp		Point layer
5.	[Port]	DX72_C3P.shp		Point layer, DJ (general port, main port)
6.	[Highway hub]	DX72_C4D.shp		Point layer
7.	[Freeway entrance]	DX72_C5H.shp		Point layer
8.	[Urban rail transit station]	DX72_C6S.shp		Point layer
9.	[Road network]	DX/DX72/GNetSet.gdb/RN0124/RN0124_ND		
		[GC control]		

7.8.3   GC Process

Table 7.32 shows the [GC Process table] of comprehensive advantage evaluation (city and county).

Table 7.32 [GC process table] of comprehensive advantage evaluation (city and county)

Step	Operation instruction	Input	Operation	Output	Description
1.	Accessibility calculation	[Central city] [Road network]	**[Description]** Accessibility analysis of transportation network **[Keywords]** {Filter field I Filter item list} %Time unit I Time list% Reclassification information **KX_ServiceAreaExt(MI30,60,90,120%5,4,3,2)**	[Accessibility of central urban area 1] DX72_QWB1	
2.	Value modification	[Accessibility of central urban area 1]	**[Description]** Raster calculation **[Keywords]** Algebraic or logical expression **KX_RasCalculator (Is Null([R1]), 1, [R1])**	[Accessibility of central urban area] DX72_QWB	1(Bad) 2(Poor) 3(Common) 4(Good) 5(Excellent)
3.	Accessibility calculation	[Airport] [Road network]	**[Description]** Accessibility analysis of transportation network **[Keywords]** {Filter field I Filter item list} %Time unit I Time list% Reclassification information **KX_ServiceAreaExt (MI60, 90,120 %5,4,3)**	[Airport service accessibility] DX72_QWC1	
4.	Accessibility calculation	[Railway station] [Road network]	**[Description]** Accessibility analysis of transportation network **[Keywords]** {Filter field I Filter item list} %Time unit I Time list% Reclassification information **KX_ServiceAreaExt (MI30, 60 %5,4)**	[Railway station accessibility] DX72_QWC2	
5.	Accessibility calculation	[Port] [Road network]	**[Description]** Accessibility analysis of transportation network **[Keywords]** {Filter field I filter item list} %Time unit I Time list% Reclassification information **KX_ServiceAreaExt (DJI Main ports%MI60, 90 %3,2)**	[Accessibility of major ports] DX72_QWC31	

(continued)

Table 7.32 (continued)

Step	Operation instruction	Input	Operation	Output	Description
6.	Accessibility calculation	[Port] [Road network]	[Description] Accessibility analysis of transportation network [Keywords] {Filter field	Filter item list} %Time unit	Time list% Reclassification information **KX_ServiceAreaExt (DJI Main ports%MI60 %2)**	[Accessibility of general ports] DX72_QWC32	
7.	Accessibility calculation	[Highway hub] [Road network]	[Description] Accessibility analysis of transportation network [Keywords] {Filter field	Filter item list} %Time unit	Time list% Reclassification information **KX_ServiceAreaExt (MI30, 60 %3,2)**	[Accessibility of highway hub] DX72_QWC4	
8.	Accessibility calculation	[Freeway entrance] [Road network]	[Description] Accessibility analysis of transportation network [Keywords] {Filter field	Filter item list} %Time unit	Time list% Reclassification information **KX_ServiceAreaExt (MI30, 60 %4,3)**	[Accessibility of Expressway access] DX72_QWC5	
9.	Accessibility calculation	[Urban rail transit station] [Road network]	[Description] Accessibility analysis of transportation network [Keywords] {Filter field	Filter item list} %Time unit	Time list% Reclassification information **KX_ServiceAreaExt (MI30, 45,60 %5,4,3)**	[Accessibility of urban rail transit stations] DX72_QWC6	

(continued)

Table 7.32 (continued)

Step	Operation instruction	Input	Operation	Output	Description
10.	Accumulated value calculation	[Airport service accessibility] [Railway station accessibility] [Accessibility of major ports] [Accessibility of general ports] [Accessibility of highway hub] [Accessibility of expressway access] [Accessibility of urban rail transit stations]	**[Description]** Raster calculation **[Keywords]** Algebraic or logical expression **KX_RasCalculator ([R1] + [R2] + [R3] + [R4] + [R5] + [R6] + [R7])**	[Accessibility of transport hub 1] DX72_QWC7	
11.	Reclassification	[Accessibility of transport hub 1]	**[Description]** Reclassification (equal distance), **[Keywords]** N1|N2|...|Nn **KX_Reclass(1|2|3|4|5)**	[Accessibility of transport hub] DX72_ QWC	1(Bad) 2(Poor) 3(Common) 4(Good) 5(Excellent)

(continued)

Table 7.32 (continued)

Step	Operation instruction	Input	Operation	Output	Description
12.	Accessibility calculation	[Road]	[**Description**] Filter + Euclidean distance + Reclassification [**Keywords**] {Operation field I Filter list} % Reclassification information **KX_SelDisReClass (DJI1 % 5: <3000I 4: 3000–6000I 1: ≥6000)**	[Accessibility of first-class highway] DX72_GL1	
13.	Accessibility calculation	[Road]	[**Description**] Filter + Euclidean distance + Reclassification [**Keywords**] {Operation field I Filter list} % Reclassification information **KX_SelDisReClass (DJI2 % 4: <3000I 3: 3000–6000I 1: ≥6000)**	[Accessibility of secondary highway] DX72_GL2	
14.	Accessibility calculation	[Highway]	[**Description**] Filter + Euclidean distance + Reclassification [**Keywords**] {Operation field I Filter list} % Reclassification information **KX_SelDisReClass (DJI3 % 3: <3000I 2: 3000–6000I 1: ≥6000)**	[Accessibility of class III highway] DX72_GL3	
15.	Accessibility calculation	[Highway]	[**Description**] Filter + Euclidean distance + Reclassification [**Keywords**] {Operation field I Filter list} % Reclassification information **KX_SelDisReClass (DJI4 % 2: <3000I 1: 3000–6000I 1: ≥6000)**	[Accessibility of class IV highway] DX72_GL4	
16.	Accumulated value calculation	[Accessibility of first-class highway] [Accessibility of secondary highway] [Accessibility of class III highway] [Accessibility of class IV highway]	[**Description**] Raster calculation [**Keywords**] Algebraic or logical expression **KX_RasCalculator ([R1] + [R2] + [R3] + [R4])**	[Accessibility of traffic trunk lines 1] DX72_QWA1	

(continued)

Table 7.32 (continued)

Step	Operation instruction	Input	Operation	Output	Description
17.	Reclassification	[Accessibility of traffic trunk lines 1]	**[Description]** Reclassification (equal distance), **[Keywords]** N1|N2|…|Nn **KX_Reclass(1|2|3|4|5)**	[Accessibility of traffic trunk lines] DX72_ QWA	1(Bad) 2(Poor) 3(Common) 4(Good) 5(Excellent)
18.	Accumulated value calculation	Accessibility of traffic trunk lines] [Accessibility of central urban area] [Accessibility of transport hub]	**[Description]** Raster calculation **[Keywords]** Algebraic or logical expression **KX_RasCalculator ([R1] + [R2] + [R3])**	[Location conditions 1] DX72_QW1	
19.	Reclassification	[Location condition 1]	**[Description]** Reclassification (equal distance), **[Keywords]** N1|N2|…|Nn **KX_Reclass(1|2|3|4|5)**	[Location condition 2] DX72_QW2	
20.	Clip [Location condition 2]	[Location condition 2] [Range layer]	**[Description]** Clip **[Keywords]** Clipped layer, Clip layer **KX_Clip**	[Location condition] DX72_QWTJP	1(Bad) 2(Poor) 3(Common) 4(Good) 5(Excellent)
21.	Kernel density calculation	[Road]	**[Description]** Kernel density calculation **[Keywords]** [Multiplication field] |Radius **KX_DENSITY2(PFfield1|1000)**	[Traffic network density 1] DX72_JTMD1	

(continued)

Table 7.32 (continued)

Step	Operation instruction	Input	Operation	Output	Description
22.	Value modification	[Traffic network density 1]	[Description] Raster calculation [Keywords] Algebraic or logical expression KX_RasCalculator ([R1] * 1000)	[Traffic network density 2] DX72_JTMD2	
23.	Reclassification	[Traffic network density 2]	[Description] Reclassification (numerical range) [Keywords] N1: <V1|N2: V1 – V2|...|Nn: ≥Vn KX_Reclass (1: <1|2: 1–3|3: 3–8|4: ≥8)	[Traffic network density 3] DX72_JTMD3	
24.	Clip	[Traffic network density 3] [Range layer]	[Description] Clip [Keywords] Clipped layer, Clip layer KX_Clip	[Traffic network density] DX72_JTMD	1(Lower) 2(Low) 3(Medium) 4(High) 5(Higher)
25.	Generate [Location advantage 1]	[Location conditions] [Traffic network density]	[Description] Judgment matrix integration [Keywords] Raster layer (row), raster layer (column) KX_Matrix ([5,4,3,2,1]|[5,4,3,2,1]|[5,5,4,3,1]|[5,5,4,2,1]|[4,4,3,1,1]|[3,3,2,1,1])	[Location advantage 1] DX72_ QWYSD1	1(Lower) 2(Low) 3(Medium) 4(High) 5(Higher)
26.	Clip [Location advantage 1]	[Location advantage 1] [Range layer]	[Description] Clip [Keywords] Clipped layer, Clip layer KX_Clip	[Location advantage] DX72_QWYSD	
27.	Make [Thematic map of location conditions]	[Location conditions]	[Description] Make thematic map [Keywords] Replace list | Background list | Resolution | {Template} | {Drawing range} KX_Mapping(C2|Boundary|200)	[Thematic map of location conditions] DX72_QWTJ.emf	

(continued)

Table 7.32 (continued)

Step	Operation instruction	Input	Operation	Output	Description
28.	Insert thematic map	[Thematic map of location conditions]	**[Description]** Insert thematic map **[Keywords]** Picture heightl {Delete content} **KX_InsertPic (12)**	1	
29.	Produce [Statistical table of location conditions]	[Statistical layer] [Location conditions]	**[Description]** Make statistical table **[Keywords]** Display field# {Unit parameter} l Counted type list l {Data filtering mode} **KX_Statistic (Namel5,4,3,2,1)**	2	
30.	Make [traffic network density thematic map]	[Traffic network density]	**[Description]** Make thematic map **[Keywords]** Replace list l Background list l Resolution l {Template} l {Drawing range} **KX_Mapping(C2l200)**	[Traffic network density thematic map] DX72_JTMD.emf	
31.	Insert thematic map	[traffic network density thematic map]	**[Description]** Insert thematic map **[Keywords]** Picture heightl {Delete content} **KX_InsertPic (12)**	3	
32.	Produce [Statistical table of traffic network density]	[Statistical layer] [Traffic network density]	**[Description]** Make statistical table **[Keywords]** Display field# {Unit parameter} l Counted type list l {Data filtering mode} **KX_Statistic(Namel5,4,3,2,1)**	4	

(continued)

Table 7.32 (continued)

Step	Operation instruction	Input	Operation	Output	Description
33.	Make [Location advantage classification map]	[Location advantage]	**[Description]** Make thematic map **[Keywords]** Replace list I Background list I Resolution I {Template} I {Drawing range} **KX_Mapping(C2IBoundaryI200)**	[Location advantage classification map] DX72_ZHYSD.emf	
34.	Insert thematic map	[Location advantage classification map]	**[Description]** Insert thematic map **[Keywords]** Picture heightI {Delete content} **KX_InsertPic (13)**	5	
35.	Produce statistical table	[Statistical layer] [Location advantage]	**[Description]** Make statistical table **[Keywords]** Display field# {Unit parameter} I Counted type list I {Data filtering mode} **KX_Statistic (NameI5,4,3,2,1)**	6	

Chapter 8
Integrated Evaluation

Integrated evaluation is an evaluation work carried out based on Individual evaluation, including the importance of ecological protection, suitability of agricultural production, suitability of urban construction, carrying scale of agricultural production and carrying scale of urban construction, etc.

8.1 Importance of Ecological Protection

8.1.1 Evaluation Method

This model refers to page 20 of *Technical Guidelines (August Edition)*.

1. Evaluation methodology

$$[\text{Importance level of ecological protection}] = f([\text{Importance of}$$
$$\text{ecosystem services}], [\text{Ecological sensitivity}],$$
$$[\text{Ecological corridor}], [\text{Trimming patch}]) \qquad (8.1)$$

2. Evaluation steps

The evaluation process is shown in Fig. 8.1.

(1) [Importance of ecosystem service] and [Ecological sensitivity] are computed according to Table 8.1 to obtain [Important level of ecological protection 1].
(2) [Important level of ecological protection 1] is modified by [Ecological corridor] to obtain [Important level of ecological protection 2].
(3) [Important level of ecological protection 2] is modified by [Trimming patch] to obtain the final [Importance level of ecological protection].

© Surveying and Mapping Press 2021
W. Zhou, *A New GeoComputation Pattern and Its Application in Dual-Evaluation*,
https://doi.org/10.1007/978-981-33-6432-5_8

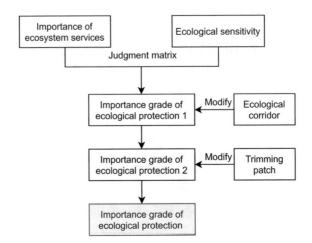

Fig. 8.1 GC flow chart of integrated evaluation of ecological protection importance

Table 8.1 Discrimination matrix of ecological protection importance level

Importance of ecosystem service function	Ecological sensitivity		
	Extremely sensitive	Sensitive	General sensitive
Very important	Very important	Very important	Very important
Important	Very important	Important	Important
Generally important	Very important	Important	Generally important

8.1.2 Input and Control

Table 8.2 shows the [Input and control table] of ecological protection importance.

Table 8.2 [Input and control table] of importance of ecological protection

Serial No.	Layer name	Physical layer	Reference page	Value and description
1.	[Importance of ecosystem services]	DX\DX1\DX1_STZYX		3(Very important)/2(Important)/ 1(Generally important)
2.	[Ecological sensitivity]	DX\DX1\DX1_STMGX		3(Extremely sensitive)/2(Sensitive)/ 1(General sensitive)
3.	[Ecological corridor]	JC1_STLD		
4.	[Trimming patch]	JC1_XBTB.shp		
[GC control]				

8.1.3   GC Process

Table 8.3 shows the [GC process table] of ecological protection importance.

Table 8.3   [GC process table] of ecological protection importance

Step	Operation instruction	Input	Operation	Output	Description
1.	Calculate [Importance grade of ecological protection 1]	[Ecological sensitivity] [Importance of ecosystem services]	**[Description]** Judgment matrix integration **[Keywords]** Raster layer (row), raster layer (column) **KX_Matrix ([3,2,1] | [3,2,1] | [3,3,3] | [3,2,2] | [3,2,1])**	[Importance grade of ecological protection 1] JC1_STZYX1	3(Very important) 2(Important) 1(Generally important)
2.	Generate [Importance grade of ecological protection 2]	[Ecological corridor] [Importance grade of ecological protection 1]	**[Description]** Raster calculation If [Ecological corridor 1] = 1, and [Importance of ecological protection 1] = 2, then [Ecological protection level] = 3 If [Ecological corridor 1] = 1, and [Importance of ecological protection 1] = 1, then [Ecological protection level] = 2 **[Keywords]** Algebraic or logical expression **KX_RasCalculator ((([R1] == 1) and ([R2] == 2), 3% ([R1] == 1) and ([R2] == 1), 2% [R2])**	[Importance grade of ecological protection 2] JC1_STZYX2	
3.	Modification	[Trimming patch] [Importance grade of ecological protection 2]	**[Description]** Raster modification **[Keywords]** Number or field **KX_MODIFYRASTER (0)**	[Importance grade of ecological protection 3] JC1_STZYX3	
4.	Clip	[Importance grade of ecological protection 3] [Range layer]	**[Description]** Clip **[Keywords]** Clipped layer, Clip layer **KX_Clip**	[Importance grade of ecological protection] JC1_STZYX	3(Very important) 2(Important) 1(Generally important)

(continued)

Table 8.3 (continued)

Step	Operation instruction	Input	Operation	Output	Description
5.	Make thematic map	[Importance grade of ecological protection]	**[Description]** Make thematic map **[Keywords]** Replace list | Background list | Resolution | {Template} | {Drawing range} **KX_Mapping (S1|Boundary|200)**	[Importance grade of ecological protection thematic map] JC1_STZYX.emf	
6.	Insert thematic map	[Importance grade of ecological protection thematic map]	**[Description]** Insert thematic map **[Keywords]** Picture height | {Delete content} KX_InsertPic (12)	1	
7.	Make statistical table	[Statistical layer] [Importance grade of ecological protection]	**[Description]** Make statistical table **[Keywords]** Display field# {Unit parameter} | Counted type list | {Data filtering mode} **KX_Statistic (Name|3,2,1)**	2	

8.1.4   Thematic Map and Statistical Table

Figure 8.2 is the thematic map of the importance of ecological protection, and Table 8.4 is a summary of the evaluation results of ecological protection importance.

8.2   Suitability of Agricultural Production

8.2.1   Evaluation Method

This model is referred to *Technical Guidelines (June Edition)*, pages 53–56.

1. **Evaluation methodology**

$$\begin{aligned}[\text{Suitability of agricultural production}] = f([\text{Soil and water}\\\text{resources foundation}], [\text{Photothermal conditions}],\\ [\text{Soil environmental capacity}], [\text{Salinization sensitivity}],\\ [\text{Meteorological disaster risk}])\end{aligned}$$

$$(8.2)$$

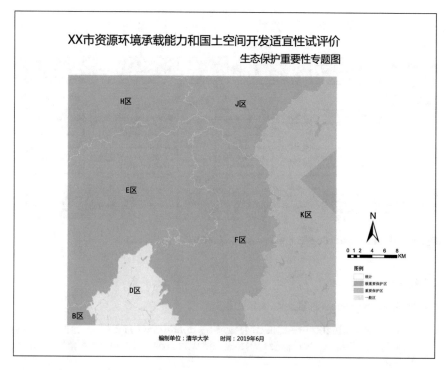

Fig. 8.2 Thematic map of the importance of ecological protection

Table 8.4 Summary of evaluation results of ecological protection importance (unit: Area, km²; proportion, %)

Region		Very important		Important		Generally important	
		Area	Ratio	Area	Ratio	Area	Ratio
* * City	B	12.56	99.9	0.00	0.0	0.02	0.2
	D	0.21	0.2	1.19	1.1	106.94	98.7
	E	454.79	100.0	0.00	0.0	0.01	0.0
	F	375.98	99.9	0.17	0.0	0.06	0.0
	H	186.73	99.9	0.00	0.0	0.00	0.0
	J	152.36	99.9	0.01	0.0	0.00	0.0
	K	38.07	9.1	379.20	90.9	0.00	0.0
	Total	1220.70	71.4	380.58	22.3	107.03	6.3

$$[\text{Soil and water resources foundation}] = f([\text{Farming conditions}],$$
$$[\text{Agricultural water supply conditions}])$$
$$(8.3)$$

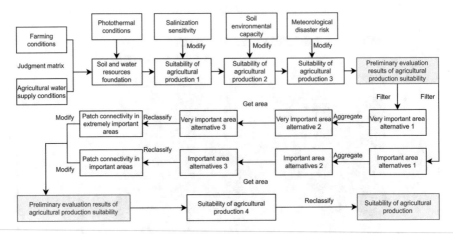

Fig. 8.3 GC flow chart of agricultural production suitability evaluation

2. **Evaluation steps**

The evaluation process is shown in Fig. 8.3.

(1) Using the evaluation results of [Farming conditions] and [Agricultural water supply conditions], compute the [Soil and water resource foundation] according to Table 8.5.

(2) Using the evaluation results of [Soil and water resources foundation] and [Photothermal conditions], the preliminary results of agricultural production grade [Suitability of agricultural production 1] is computed according to Table 8.6.

(3) Modify [Suitability of agricultural production 1] by using [Salinization sensitivity]. For areas with high salinization sensitivity, the preliminary evaluation result [Suitability of agricultural production 1] is reduced by one grade to obtain [Suitability of agricultural production 2].

(4) Use [Soil environmental capacity] to modify [Suitability of agricultural production 2]. If the evaluation result of soil environmental capacity is the lowest,

Table 8.5 Discrimination matrix of soil and water resources foundation directed by agricultural function

Agricultural water supply conditions	Farming conditions				
	Higher	High	Common	Low	Lower
Excellent	Excellent	Excellent	Good	Common	Bad
Good	Excellent	Excellent	Good	Poor	Bad
Common	Excellent	Good	Common	Poor	Bad
Poor	Good	Common	Poor	Bad	Bad
Bad	Bad	Bad	Bad	Bad	Bad

Table 8.6 Preliminary suitability rating of agricultural production function direction

Photothermal conditions	Soil and water resource foundation				
	Excellent	Good	Common	Poor	Not suitable
Excellent	Suitable	Suitable	More suitable	Generally suitable	Not suitable
Good	Suitable	More suitable	More suitable	Less suitable	Not suitable
Common	Suitable	More suitable	Generally suitable	Less suitable	Not suitable
Poor	More suitable	Generally suitable	Less suitable	Not suitable	Not suitable
Bad	Not suitable	Not suitable	Not suitable	Not suitable	Not suitable

Table 8.7 Reference thresholds for classification of patch connectivity evaluation (unit: mu)

Patch connectivity	Lower	Low	Common	High	Higher
Plain field area	<150	150–400	400–600	600–900	≥900
Mountainous and hilly land area	<80	80–150	150–250	250–400	≥400

[Suitability of agricultural production 2] is reduced by two levels to obtain [Suitability of agricultural production 3].

(5) Modify [Suitability of agricultural production 3] by using [Meteorological disaster risk]. For areas with high risk of meteorological disasters, [Suitability of agricultural production 3] is adjusted to a more suitable grade, and the [Preliminary evaluation results of agricultural production suitability] is obtained.

(6) The [Patch connectivity] is obtained through aggregation operation by using the [Preliminary evaluation results of agricultural production suitability]. See Table 8.7 for the evaluation and classification method of patch connectivity.

(7) The [Preliminary evaluation results of agricultural production suitability] was modified by using the [Patch connectivity] to obtain the evaluation result of [Suitability of agricultural production 4] of 5 grades.

(8) The high and higher [Suitability of agricultural production 4] is designated as suitable, the general and lower [Suitability of agricultural production 4] is designated as generally suitable, and the low [Suitability of agricultural production 4] is designated as unsuitable, thus obtaining the final evaluation results [Suitability of agricultural production].

8.2.2 Input and Control

Table 8.8 is the [Input and control table] of agricultural production suitability.

Table 8.8 Agricultural production suitability [Input and control table]

Serial No.	Layer name	Physical layer	Reference page	Value and description
1.	[Farming conditions]	DX\DX2\DX2_NYGZTJ		5(Excellent)/4(Good)/3(Common)/2(Poor)/1(Bad)
2.	[Agricultural water supply conditions]	DX\DX3\DX3_NYGSTJ		5(Excellent)/4(Good)/3(Common)/2(Poor)/1(Bad)
3.	[Photothermal conditions]	DX\DX4\DX4_GRTJ		5(Excellent)/4(Good)/3(Common)/2(Poor)/1(Bad)
4.	[Salinization sensitivity]	DX\DX1\DX1_YZHMGX		5(Extremely sensitive)/4(Highly sensitive) / 3 (Moderately sensitive) / 2 (Mildly sensitive) / 1 (Insensitive)
5.	[Soil environmental capacity]	DX\DX5\DX5_TRHJRL		5(Excellent)/4(Good)/3(Common)/2(Poor)/1(Bad)
6.	[Meteorological disaster risk]	DX\DX6\DX6_QXZHFX		5(Higher)/4(High)/3(Medium)/2(Low)/1(Lower)
	[GC control]			

8.2.3 GC Process

Table 8.9 shows the [GC process table] of agricultural production suitability.

8.2.4 Thematic Map and Statistical Table

Figure 8.4 is the thematic map of agricultural production suitability, and Table 8.10 is the summary table of agricultural production suitability evaluation results.

8.3 Suitability of Urban Construction

8.3.1 Evaluation Method

This model is referred to pages 56–58 in *Technical Guidelines (June Edition)*.

1. **Evaluation methodology**

$$[\text{Suitability grade of urban construction}] = f([\text{Soil and water resource}$$
$$\text{foundation}], [\text{Water and air environmental capacity}],$$
$$[\text{Comfort level}], [\text{Location advantage degree}]) \qquad (8.4)$$

$$[\text{Soil and water resource foundation}] = f([\text{Urban construction conditions}],$$
$$[\text{Urban water supply conditions}]) \quad (8.5)$$

$$[\text{Water and air environmental capacity}] = f([\text{Water environmental capacity}],$$
$$[\text{Atmospheric environmental capacity}])$$
$$(8.6)$$

2. **Evaluation steps**

The evaluation process is shown in Fig. 8.5.

(1) Use [Urban construction conditions] and [Urban water supply conditions] to compute [Soil and water resource foundation] according to Table 8.11.
(2) Use [Water environmental capacity] and [Atmospheric environmental capacity] to modify [Soil and water resource foundation], [Suitability of urban construction 1] was obtained.

Table 8.9 [GC process table] of agricultural production suitability

Step	Operation instruction	Input layer	Operation	Output layer and naming	Description
1.	Judgment matrix integration	[Farming conditions] [Agricultural water supply conditions]	**[Description]** Judgment matrix integration **[Keywords]** Raster layer (row), Raster layer (column) **KX_Matrix** ([5,4,3,2,1]|[5,5,4,3,1]|[5,5,4,2,1]|[5,4,3,2,1]|[4,3,2,1,1]|[1,1,1,1,1])	[Soil and water resources foundation] JC2_STZYJC	5(Excellent) 4(Good) 3(Common) 2(Poor) 1(Bad)
2.	Judgment matrix integration	[Soil and water resource foundation] [Photothermal conditions]	**[Description]** Judgment matrix integration **[Keywords]** Raster layer (row), Raster layer (column) **KX_Matrix** ([5,4,3,2,1]|[5,5,4,3,1]|[5,5,4,3,1]|[5,4,4,2,1]|[4,3,2,1,1]|[1,1,1,1,1])	[Suitability of agricultural production 1] JC2_NYSYX1	5(Suitable) 4(More suitable) 3(Generally suitable) 2(Less suitable) 1(Not suitable)
3.	Modification operation	[Salinization sensitivity] [Suitability of agricultural production 1]	**[Description]** Raster calculation [Salinization sensitivity] = 5, [Suitability of agricultural production 1] = [Suitability of agricultural production 1] − 1 **[Keywords]** Algebraic or logical expression **KX_RasCalculator** ([R2] == 5, [R2] − 1% [R2])	[Suitability of agricultural production 2] JC2_NYSYX2	

(continued)

Table 8.9 (continued)

Step	Operation instruction	Input layer	Operation	Output layer and naming	Description
4.	Modification operation	[Soil environmental capacity] [Suitability of agricultural production 2]	**[Description]** Raster calculation [Soil environmental capacity] = 1, [Suitability of agricultural production 3] = [Suitability of agricultural production 2] − 2 **[Keywords]** Algebraic or logical expression **KX_RasCalculator (([R2] == 5), [R2] − 1%[R2])**	[Suitability of agricultural production 3] JC2_NYSYX3	
5.	Modification operation	[Meteorological disaster risk] [Suitability of agricultural production 3]	**[Description]** Raster calculation [Meteorological disaster risk] = 5 and [Suitability of agricultural production 3] = 5, [Suitability of agricultural production 4] = 4 **[Keywords]** Algebraic or logical expression **KX_RasCalculator (([R1] == 5) and (([R2] == 5), 4%[R2])**	[Suitability of agricultural production 4] JC2_NYSYX4	
6.	Value modification	[Suitability of agricultural production 4]	**[Description]** Raster calculation **[Keywords]** Algebraic or logical expression **KX_RasCalculator (Con(([R1] == 0), 1, [R1]))**	[Suitability of agricultural production 5] JC2_NYSYX5	
7.	Clip	[Suitability of agricultural production 5] [Range layer]	**[Description]** Clip **[Keywords]** Clipped layer, Clip layer **KX_Clip**	[Suitability of agricultural production 6] JC2_NYSYX6	
8.	Select the most important area	[Suitability of agricultural production 6]	**[Description]** Raster calculation Select [Suitability of agricultural production 6] = 5, 4 **[Keywords]** Algebraic or logical expression **KX_RasCalculator (([R1] == 5) or ([R1] == 4), 1)**	[Very important area alternative 1] JC2_JZYQBX1	

(continued)

Table 8.9 (continued)

Step	Operation instruction	Input layer	Operation	Output layer and naming	Description
9.	Aggregation operation	[Very important area alternative 1]	[Keywords] Raster aggregation [Keywords] Aggregation distance **KX_Aggregate_Ras (1)**	[Very important area alternative 2] JC2_JZYQBX2	
10.	Patch area calculation	[Very important area alternative 2]	[Description] Patch area calculation [Keywords] Unit **KX_GetArea_Ras (MU)**	[Very important area alternative 3] JC2_JZYQBX3	
11.	Patch concentration calculation	[Very important area alternative 3]	[Description] Reclassification (numerical range) [Keywords] N1: <V1|N2: V1 − V2|…|Nn: ≥Vn **KX_Reclass (1: <150|2: 150−400|3: 400−600|4: 600−900|5: ≥900)**	[Very important area alternative 4] JC2_JZYQBX4	Adopt the standard of plain area
12.	Select important area	[Suitability of agricultural production 6]	[Description] Raster calculation Select [Agricultural carrying capacity] = 3, 4, 5 [Keywords] Algebraic or logical expression **KX_RasCalculator (([R1] == 3) or ([R1] == 4) or ([R1] == 5), 1)**	[Important area alternative 1] JC2_ZYQBX1	
13.	Aggregation operation	[Important area alternative 1]	[Keywords] Raster aggregation [Keywords] Aggregation distance **KX_Aggregate_Ras (1)**	[Important area alternative 2] JC2_ZYQBX2	
14.	Patch area calculation	[Important area alternative 2]	[Description] Patch area calculation [Keywords] Unit **KX_GetArea_Ras (mu)**	[Important area alternative 3] JC2_ZYQBX3	
15.	Patch concentration calculation	[Important area alternative 3]	[Description] Reclassification (numerical range) [Keywords] N1: <V1|N2: V1 − V2|…|Nn: ≥Vn **KX_Reclass (1: <150|2: 150−400|3: 400−600|4: 600−900|5: ≥900)**	[Important area alternative 4] JC2_ZYQBX4	

(continued)

Table 8.9 (continued)

Step	Operation instruction	Input layer	Operation	Output layer and naming	Description
16.	High zone modification	[Suitability of agricultural production 4] [Very important area alternative 4]	**[Description]** Judgment matrix integration **[Keywords]** Raster layer (row), raster layer (column) **KX_Matrix** **([5,4,3,2,1]||[5,4,3,2,1]||[5,4,4,4,4]||[4,4,3,3,3]||[6,6,6,6,6]||[2,2,2,2,2]||[1,1,1,1,1])**	[Suitability of agricultural production 7] JC2_NYSYX7	
17.	Medium area modification	[Suitability of agricultural production 7] [Important area alternatives 4]	**[Description]** Raster calcualtion **[Keywords]** Algebraic or logical expression **KX_RasCalculator (([R1] == 6) and ([R2] == 5), 4%([R1] == 6) and ([R2] == 4), 3%([R1] == 6) and ([R2] == 4), 3%([R1] == 6) and ([R2] == 2), 2%([R1] == 6) and ([R2] == 2), 2%[R1])**	[Suitability of agricultural production 8] JC2_NYSYX8	
18.	Convert level 5 to level 3	[Suitability of agricultural production 8]	**[Description]** Reclassification (class modification) **[Keywords]** N1:C1, C2, ...lN2:C3, C4, ...l.... **KX_Reclass (3: 5,4|2: 3,2|1: 1)**	[Suitability of agricultural production 9] JC2_NYSYX9	
19.	Clip	[Suitability of agricultural production 9] [Range layer]	**[Description]** Clip **[Keywords]** Clipped layer, Clip layer **KX_Clip**	[Suitability of agricultural production] JC2_NYSYX	3(Suitable) 2(Generally suitable) 1(Not suitable)

(continued)

Table 8.9 (continued)

Step	Operation instruction	Input layer	Operation	Output layer and naming	Description
20.	Make thematic map	[Suitability of agricultural production]	**[Description]** Make thematic map **[Keywords]** Replace list | Background list |Resolution | {Template} | {Drawing range} **KX_Mapping (S2|Boundary|200)**	[Thematic map of agricultural production suitability] JC2_NYSYX.EMF 1	
21.	Insert thematic map	[Thematic map of agricultural production suitability]	**[Description]** Insert thematic map **[Keywords]** Picture height | {Delete content} **KX_InsertPic (12)**		
22.	Make statistical table	[Statistical layer] [Suitability of agricultural production]	**[Description]** Make statistical table **[Keywords]** Display field# {Unit parameter} |Counted type list | {Data filtering mode} **KX_Statistic (Name|3,2,1)**	2	

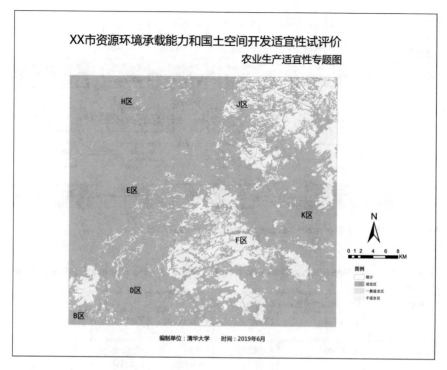

Fig. 8.4   Thematic map of agricultural production suitability

Table 8.10 Summary of agricultural production suitability evaluation results (unit: area, km^2; proportion, %)

Region		Suitable		Generally suitable		Not suitable	
		Area	Ratio	Area	Ratio	Area	Ratio
* * City	B	11.75	93.4	0.00	0.0	0.83	6.6
	D	102.92	95.0	0.00	0.0	5.42	5.0
	E	408.46	89.8	0.00	0.0	46.34	10.2
	F	259.27	68.9	0.00	0.0	116.95	31.1
	H	184.54	98.8	0.00	0.0	2.19	1.2
	J	104.02	68.2	0.00	0.0	48.35	31.7
	K	310.18	74.4	0.00	0.0	107.09	25.7
	Total	1381.14	80.8	0.00	0.0	327.17	19.1

(3) [Suitability of urban construction 1] is modified by [Comfort]. For comfort level, if the index level is the lowest, the preliminary evaluation result of urban suitability decreases by one level to obtain [Suitability of urban construction 2].

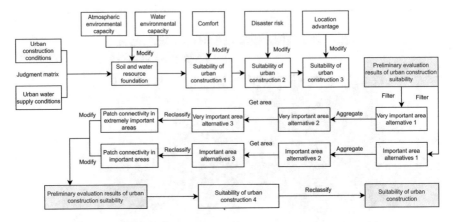

Fig. 8.5 GC flow chart of urban construction suitability evaluation

Table 8.11 Basic reference discrimination matrix of water and soil resource directed by urban construction function

Urban water supply conditions	Conditions for urban construction				
	Excellent	Good	Common	Poor	Bad
Excellent	Suitable	Suitable	More suitable	Generally suitable	Not suitable
Good	Suitable	Suitable	More suitable	Less suitable	Not suitable
Common	Suitable	More suitable	Generally suitable	Less suitable	Not suitable
Poor	More suitable	More suitable	Generally suitable	Not suitable	Not suitable
Bad	Generally suitable	Generally suitable	Less suitable	Not suitable	Bad

(4) Modify [Suitability of urban construction 2] by using [Disaster risk]. The preliminary evaluation results of urban suitability are suitable and more suitable, but the land space with high disaster risk is adjusted to be generally suitable. For the land space where the preliminary evaluation results are suitable but the disaster risk is relatively high, adjust it to be more suitable and obtain [Suitability of urban construction 3].

(5) [Suitability of urban construction 3] is modified by using [Location advantage degree]. If the [Location advantage degree] is the lowest value, the suitability level of urban construction will be directly determined as inappropriate; If the result is poor, the initial result will be lowered by one level; If the results are

Table 8.12 Grading reference thresholds for patch connectivity evaluation (unit: km^2)

Patch connectivity	Lower	Low	Medium	High	Higher
Patch area	<0.25	0.25–0.5	0.5–1.0	1.0–2.0	≥2.0

good, the less suitable, generally suitable and more suitable files in the preliminary planning results will be raised by one level respectively to obtain the [Preliminary evaluation results of urban construction suitability].

(6) The [Patch connectivity] is obtained through aggregation operation by using the [Preliminary evaluation results of urban construction suitability]. See Table 8.12 for the evaluation and classification method of patch connectivity.

(7) The [Preliminary evaluation results of urban construction suitability] are modified by using the [Patch connectivity] to obtain 5-level evaluation results of [Suitability of urban construction 4].

(8) The high and higher results in [Suitability of urban construction 4] is designated as suitable, the general and lower results in [Suitability of urban construction 4] is designated as generally suitable, and the lower results in [Suitability of urban construction 4] is designated as inappropriate, thus obtaining the final evaluation results of [Suitability of urban construction].

8.3.2 Input and Control

Table 8.13 shows the [Input and control table] of urban construction suitability.

8.3.3 GC Process

Table 8.14 is the [GC process table] of the suitability of urban construction.

8.3.4 Thematic Map and Statistical Table

Figure 8.6 is the thematic map of urban construction suitability, and Table 8.15 is a summary of the evaluation results of urban bearing grade.

Table 8.13 [Input and control] of suitability for urban construction

Serial No.	Layer name	Physical layer	Reference page	Value and description
1.	[Urban construction conditions]	DX\DX2\DX2_CZJSTJ		5(Excellent)/4(Good)/ 3(Medium)/2(Poor)/1(Bad)
2.	[Urban water supply conditions]	DX\DX3\ DX3_CZGSTJ		5(Excellent)/4(Good)/ 3(Medium)/2(Poor)/1(Bad)
3.	[Water environmental capacity]	DX\DX5\DX5_SHJRL		5(Higher)/4(High)/3(Medium)/ 2(Low)/1(Lower)
4.	[Atmospheric environmental capacity]	DX\DX5\DX5_DQHJRL		5(Higher)/4(High)/3(Medium)/ 2(Low)/1(Lower)
5.	[Comfort]	DX\DX4\DX4_SSD		5(Excellent)/4(Good)/ 3(Medium)/2(Poor)/1(Bad)
6.	[Disaster risk]	DX\DX6\DX6_ZHWXX		4(Higher)/3(High)/3(Medium)/ 2(Medium)/1(Lower)
7.	[Location advantage degree]	DX\DX72\DX72_QWY SD		5(Excellent)/4(Good)/ 3(Medium)/2(Poor)/1(Bad)
		[GC control]		

Table 8.14 [GC process table] of urban construction suitability

Step	Step description	Input	Operation	Output layer	Description
1.	Judgment matrix integration	[Urban construction conditions] [Urban water supply conditions]	**[Description]** Judgment matrix integration **[Keywords]** Raster layer (row), Raster layer (column) **KX_Matrix** ([5,4,3,2,1]‖[5,4,3,2,1]‖[5,5,4,3,1]‖[5,5,4,2,1]‖[5,4,3,2,1]‖[4,4,3,1,1]‖[3,3,2,1,1])	[Soil and water resources foundation] **JC3_STZYJC**	5(Suitable) 4(More suitable) 3(Generally suitable) 2(Less suitable) 1(Not suitable)
2.	Modification operation	[Soil and water resources foundation] [Water environmental capacity] [Atmospheric environmental capacity]	**[Description]** Raster calculation If [Water environmental capacity] == 1 and [Atmospheric environmental capacity] == 1, then [Suitability of urban construction 1] = [Soil and water resources foundation] − 2 If [Water environmental capacity] == 1 or [Atmospheric environmental capacity] == 1, [Suitability of urban construction 1] = [Soil and water resources foundation] − 1 **[Keywords]** Algebraic or logical expression **KX_RasCalculator (([R2] == 1) and ([R3] == 1), [R1] − 2%([R2] == 1) or ([R3] == 1), [R1] − 1%[R1])**	[Suitability of urban construction 1] JC3_CZSYX1	
3.	Value modification	[Suitability of urban construction 1]	**[Description]** Raster calculation **[Keywords]** Algebraic or logical expression **KX_RasCalculator (([R1] < 1), 1% [R1])**	[Suitability of urban construction 2] **JC3_CZSYX2**	

(continued)

Table 8.14 (continued)

Step	Step description	Input	Operation	Output layer	Description
4.	Modification operation	[Comfort] [Suitability of urban construction 2]	[Description] Raster calculation If [Comfort] = 1, [Suitability of urban construction 3] = [Suitability of urban construction 2] − 1 [Keywords] Algebraic or logical expression KX_RasCalculator (([R1] == 1), [R2] − 1%[R2])	[Suitability of urban construction 3] **JC3_CZSYX3**	
5.	Modification operation	[Disaster risk] [Suitability of urban construction 3]	[Description] Raster calculation If [Disaster risk] = 4 and ([Suitability of urban construction 3] = 5 or [Suitability of urban construction 3] = 4), 3 If [Disaster risk] = 3 and [Suitability of urban construction 3] = 4, 4 [Keywords] Algebraic or logical expression KX_RasCalculator (([R1] == 4) and (([R2] == 5) or ([R2] == 4)), 4%([R1] == 3) and ([R2] == 5), 4%[R2])	[Suitability of urban construction 4] **JC3_CZSYX4**	
6.	Modification operation	[Location advantage degree] [Suitability of urban construction 3]	[Description] Raster calculation If [Location advantage degree] = 1, then [Suitability of urban construction 4] = 1 If [Location advantage degree] = 2, then [Suitability of urban construction 4] = [Suitability of urban construction 3] − 1 If [Location advantage degree] = 5 and ([Suitability of urban construction 4] = [Suitability of urban construction 3] + 1) [Keywords] Algebraic or logical expression KX_RasCalculator (([R1] == 1), 1%([R1] == 2), [R2] − 1%([R1] ==5), [R2] + 1%[R2])	[Suitability of urban construction 5] **JC3_CZSYX5**	
7.	Value modification	[Suitability of urban construction 5]	[Description] Raster calculation [Keywords] Algebraic or logical expression KX_RasCalculator (([R1] < 1), 1%([R1] > 5), 5%[R1])	[Suitability of urban construction 6] **JC3_CZSYX6**	

(continued)

Table 8.14 (continued)

Step	Step description	Input	Operation	Output layer	Description
8.	Clip	[Suitability of urban construction 6] [Range layer]	**[Description]** Clip **[Keywords]** Clipped layer, Clip layer **KX_Clip**	[Suitability of urban construction 7] **JC3_CZSYX7**	
9.	Select the most important area	[Suitability of urban construction 7]	**[Description]** Raster calculation Select [Suitability of urban construction 7] = 5, 4 **[Keywords]** Algebraic or logical expression **KX_RasCalculator (([R1] == 5) or ([R1] == 4), 1)**	[Very important area alternative 1] **JC3_JZYQ BX1**	
10.	Aggregation operation	[Very important area alternative 1]	**[Description]** Raster aggregation **[Keywords]** Aggregation distance **KX_Aggregate_Ras (1)**	[Very important area alternative 2] **JC3_JZYQBX2**	
11.	Patch area calculation	[Very important area alternative 2]	**[Description]** Patch area calculation **[Keywords]** Unit **KX_GetArea_Ras (km)**	[Very important area alternative 3] JC3_JZYQBX3	
12.	Patch concentration calculation	[Very important area alternative 3]	**[Description]** Reclassification (numerical range) **[Keywords]** N1: <V1|N2: V1 − V2|...|Nn: ≥Vn **KX_Reclass (1: <0.25|2: 0.25–0.5|3: 0.5–1.0|4: 1–2|5: ≥2)**	[Very important area alternative 4] **JC3_JZYQBX4**	
13.	Select important areas	[Suitability of urban construction 7]	**[Description]** Raster calculation Select [Suitability of urban construction 7] = 3, 4, 5 **[Keywords]** Algebraic or logical expression **KX_RasCalculator (([R1] == 3) or ([R1] == 4) or ([R1] == 5), 1)**	[Important area alternative 1] **JC3_ZYQBX1**	
14.	Aggregation operation	[Important area alternatives 1]	**[Description]** Raster aggregation **[Keywords]** Aggregation distance **KX_Aggregate_Ras (1)**	[Important area alternative 2] **JC3_ZYQBX2**	

(continued)

Table 8.14 (continued)

Step	Step description	Input	Operation	Output layer	Description
15.	Patch area calculation	[Important area alternative 2]	[Description] Patch area calculation [Keywords] Unit **KX_GetArea_Ras (km)**	[Important area alternative 3] **JC3_ZYQBX3**	
16.	Patch concentration calculation	[Important area alternative 3]	[Description] Reclassification (numerical range) [Keywords] Class N1: <V1|N2: V1 − V2|...|Nn: ≥ Vn **KX_Reclass (1: <0.25|2: 0.25–0.5|3:0.5–1.0|4:1–2 |5: ≥ 2)**	[Important area alternative 4] JC3_ZYQBX4	
17.	Higher and high area modification	[Suitability of urban construction 1] [Very important area alternative 4]	[Description] Judgment matrix integration [Keywords] Raster layer (row), raster layer (column) **KX_Matrix ([5,4,3,2,1||5,4,3,2,1||5,4,4,4,4||4,4,3,3,3||6,6,6,6,6||2,2,2,2,2||[1,1,1,1,1])**	[Suitability of urban construction 8] JC3_CZSYX8	
18.	Medium area modification	[Suitability of urban construction 8] [Important area alternative 4]	[Description] Raster calculation [Keywords] Algebraic or logical expression **KX_RasCalculator (([R1] == 6) and ([R2] == 5), 3%([R1] == 6) and ([R2] == 4), 3%([R1] == 6) and ([R2] == 4), 3%([R1] == 6) and ([R2] == 2), 2%([R1] == 6) and ([R2] == 2), 2%[R1])**	[Suitability of urban construction 9] JC3_CZSYX9	
19.	Convert level 5 to level 3	[Suitability of urban construction 9]	[Description] Reclassification (class modification) [Keywords] N1: C1, C2, ...|N2: C3, C4, ...|... **KX_Reclass (3:5,4|2: 3,2|1:1)**	[Suitability of urban construction 10] **JC3_CZSYX10**	3(Suitable) 2(Generally suitable) 1(Not suitable)

(continued)

Table 8.14 (continued)

Step	Step description	Input	Operation	Output layer	Description
20.	Clip	[Suitability of urban construction 10] [Range layer]	**[Description]** Clip **[Keywords]** Clipped layer, Clip layer **KX_Clip**	[Suitability of urban construction] **JC3_CZSYX**	
21.	Make thematic map	[Suitability of urban construction]	**[Description]** Make thematic map **[Keywords]** Replace list l Background list lResolution l {Template} l {Drawing range} **KX_Mapping (S3lBoundaryl200)**	[Thematic map of urban construction suitability] **JC3_CZSYX.EMF**	
22.	Insert thematic map	[Thematic map of urban construction suitability]	**[Description]** Insert thematic map **[Keywords]** Picture heightl {Delete content} **KX_InsertPic (12)**	1	
23.	Make statistical table	[Statistical layer] [Suitability of urban construction]	**[Description]** Make statistical table **[Keywords]** Display field# {Unit parameter} l Counted type list l {Data filtering mode} **KX_Statistic (Namel3,2,1)**	2	

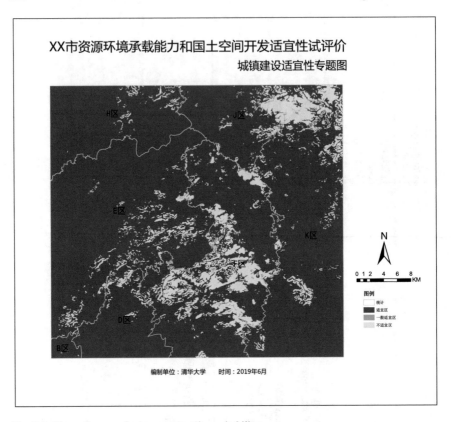

Fig. 8.6 Thematic map of urban construction suitability

Table 8.15 Summary of urban bearing grade evaluation results (unit: area, km^2; proportion, %)

Region		Suitable		Generally suitable		Not suitable	
		Area	Ratio	Area	Ratio	Area	Ratio
* * **City**	B	12.49	99.3	0.00	0.0	0.09	0.7
	D	104.11	96.1	0.00	0.0	4.23	3.9
	E	429.65	94.5	0.06	0.0	25.09	5.5
	F	286.20	76.1	1.13	0.3	88.88	23.6
	H	182.13	97.5	0.00	0.0	4.59	2.5
	J	123.84	81.2	0.63	0.4	27.91	18.3
	K	376.47	90.3	0.24	0.1	40.57	9.7
	Total	1514.89	88.7	2.06	0.1	191.35	11.2

Chapter 9
Optional Evaluation

Optional evaluation is the evaluation content carried out in all selected places according to the actual situation in various places, mainly including the suitability evaluation of marine development and utilization, the importance of cultural protection and the suitability evaluation of mineral resources development.

9.1 Suitability of Marine Development and Utilization

9.1.1 Evaluation Method

This model refers to pages 63–68 in *Technical Guidelines (June Edition)*. This chapter only introduces the suitability evaluation model of port construction.

1. **Evaluation methodology**

$$[\text{Suitability of port construction}] = f([\text{Conditions of port construction}],$$
$$[\text{Annual average risk index of storm surge}], [\text{Location advantage degree}]) \tag{9.1}$$

$$[\text{Conditions of port construction}] = f([\text{Utilization conditions of}$$
$$\text{shoreline resource}], [\text{Urban construction conditions}]) \tag{9.2}$$

$$[\text{Utilization conditions of shoreline resource}] = f([\text{Types of sediment}$$
$$\text{along shoreline}], [10\,\text{m isobath}]) \tag{9.3}$$

© Surveying and Mapping Press 2021
W. Zhou, *A New GeoComputation Pattern and Its Application in Dual-Evaluation*,
https://doi.org/10.1007/978-981-33-6432-5_9

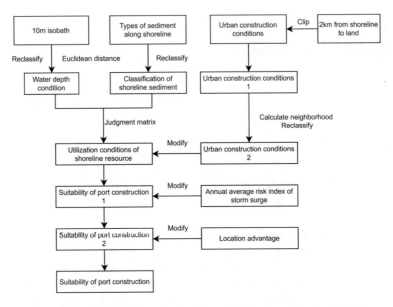

Fig. 9.1 GC flow chart of suitability evaluation for marine development and utilization

2. **Evaluation steps**

The evaluation process is shown in Fig. 9.1.

(1)   The [Types of sediment along shoreline] are reclassified and rasterized according to Table 9.1 to obtain [Classification of shoreline sediment].

(2)   After Euclidean distance operation is carried out on the [10 m isobath], it is reclassified according to Table 9.2 to obtain [Water depth condition].

(3)   Calculate with [Classification of shoreline sediment] and [water depth condition] according to Table 9.3 to obtain [Utilization conditions of shoreline resource].

(4)   The [Urban construction conditions] is clipped out from the [2 km from shoreline to land] to obtain [Urban construction conditions 1].

(5)   Neighborhood analysis is carried out on [Urban construction conditions 1], and [Urban construction conditions 2] is obtained by taking the average value.

(6)   Modify [Urban construction conditions 2] to [Utilization conditions of Shoreline Resource]. For areas with low results in [Urban construction conditions 2], the [Utilization conditions of Shoreline Resource] will be reduced by two levels as the [Port construction condition grade]; For areas with lower results in [Urban construction conditions 2], the [Utilization conditions of Shoreline Resource] will be lowered by one level as the [Port construction condition grade].

(7)   The [Annual average risk index of storm surge] is classified according to Table 9.4 to obtain the [marine disaster risk].

Table 9.1   Classification of shoreline sediment types

Evaluation grade	Higher	Medium	Lower
Sediment type of coastline	Sandy coast	Muddy coast	Bedrock coast

Table 9.2   Classification table of shortest distance from 10 m isobath (unit: km)

Evaluation grade	Excellent	Good	Medium	Poor	Bad
The shortest distance to 10 m isobath	≤1.5	1.5–3	3–4.5	4.5–6	>6

Table 9.3   Reference discrimination matrix of shoreline resources utilization conditions

Water depth conditions	Sediment type of coastline		
	Higher	Medium	Lower
Excellent	Higher	High	Medium
Good	High	Medium	Low
Common	Medium	Low	Lower
Poor	Low	Lower	Lower
Bad	Lower	Lower	Lower

Table 9.4   Average annual risk classification of storm surge

Risk level	Excellent	Good	Common	Bad
Annual average risk index of storm surge	≤2.0	2.0–3.5	3.5–7.0	>7.0

(8)   Modify the [Port construction condition grade] with [Marine disaster risk] to obtain the [Suitability of port construction]. For areas with high and higher results in [Port construction condition grade] but high [Marine disaster risk], the suitability level will be adjusted to medium. For regions with high results in [Port construction condition grade] and high [Marine disaster risk], the suitability level will be adjusted to higher.

(9)   Modify the initial judgment result of [Port construction condition grade] with [Location Advantage Degree] to obtain the final result of [Suitability of port construction]; For regions with low [Location advantage degree], the grade of [Suitability of port construction] will be adjusted to low; For areas with lower results, the [Suitability of port construction] will be lowered by one level.

9.1.2   Input and Control

Table 9.5 is the [Input and control table] for the suitability of marine development and utilization.

Table 9.5 [Input and control table] of suitability for marine development and utilization

Serial No.	Layer name	Physical layer	Reference page	Value and description
1.	[Type of Sediment along coastline]	KX1_AXZDLX.shp		LX: Sandy coast, Muddy coast, Bedrock coast
2.	[Urban construction conditions]	KX1_CZJSTJ		5(Higher)/4(High)/3(Medium)/ 2(Low)/1(Lower)
3.	[Annual average risk index of storm surge]	KX1_FBCZS		
4.	[Location advantage degree]	KX1_QWYSD		5(Excellent)/4(Good)/3(Common)/ 2(Poor)/1(Bad)
5.	[10 m isobath]	KX1_10DSX.shp		Polyline layer
6.	[2 km from shoreline to land]	KX1_2km .shp		Polygon layer
[GC control]				

9.1.3   GC Process

Table 9.6 shows the [GC Process table] of the suitability of marine development and utilization.

9.2   The Importance of Cultural Protection

9.2.1   Evaluation Method

This model is referred to *Technical Guidelines (June Edition)*, pages 69–73.

1. **Evaluation methodology**

$$[\text{Importance level of cultural protection}] = f([\text{Legal cultural protection space}],$$
$$[\text{Potential cultural protection space}], \ [\text{Contact space of cultural protection}],$$
$$[\text{Cultural resources gathering area}]) \tag{9.4}$$

$$[\text{Potential cultural protection space}] = f([\text{Potential historical area}],$$
$$[\text{Historical traffic routes}]) \tag{9.5}$$

Table 9.6 [GC process table] of suitability of marine development and utilization

Step	Operation instruction	Input	Operation	Output	Description
1.	Reclassification	[Type of sediment along coastline]	**[Description]** Vector reclassification **[Keywords]** Source field, Destination field, Reclassification field or Default value # New Category: Old Category 1, Old Category 2, … **KX_Reclass_Vec (LX,0#1: Bedrock coast|2: Muddy coast|3: Sandy coast)**	[Type of sediment along coastline R] **KX1_AXZDLX**	3(High) 2(Medium) 1(Low)
2.	Calculate water depth conditions	[10 m isobath]	**[Description]** Filter – Euclidean distance – Reclassification **[Keywords]** Filter field | Filter condition % Reclassification information **KX_SelDisReClass (5: <1500|4: 1500–3000| 3: 3000–4500 |2: 4500–6000 | 1: ≥6000)**	[Water depth condition] KX1_SSTJ	5(Excellent) 4(Good) 3(Common) 2(Poor) 1(Bad)
3.	Judgment matrix integration	[Type of sediment along coastline] [Water depth condition]	**[Description]** Judgment matrix integration **[Keywords]** Polyline description |Column description| Polyline 1 description |Polyline 2 description|… **KX_Matrix([3,2,1]|[5,4,3,2,1]||[5,4,3]|[4,3,2]||[3,2,1]|[2,1,1]|[1,1,1])**	[Utilization conditions of shoreline resources] **KX1_ZYLYTJ**	5(Higher) 4(High) 3(Medium) 2(Low) 1(Lower)
4.	Clip	[Urban construction conditions] [Range layer]	**[Description]** Clip **[Keywords]** Clipped layer, clip layer **KX_ExtractByMask**	[Urban construction conditions 1] **KX1_CZJSTJ1**	
5.	Regional analysis	[2 km from shoreline to land] [Urban construction conditions]	**[Description]** Factor calculator (buffer) **[Keywords]** Target field# Statistic item (Min, Max, Mean, Sum) or multiplication field | Standardization (0-not processed, 1-normalized, 2-denormalized | buffer range) **KX_EF_Buffer (FM#Mean|0|0)**	[2 km from shoreline to land 1] **KX1_2km.shp**	

(continued)

Table 9.6 (continued)

Step	Operation instruction	Input	Operation	Output	Description
6.	Rasterization	[2 km from shoreline to land 1]	[Description] Rasterization [Keywords] Operation field |Grid size **KX_FeatureToRaster (FM)**	[Integration of urban construction conditions 1] KX1_JSTJJC1	
7.	Reclassification	[Integration of urban construction conditions 1]	[Description] Reclassification (numerical range) [Keywords] C1: <V1|C2: V1 − V2|...|Cn: ≥Vn **KX_Reclass (1: <2|2: 2−3| 3: 3−4 |4: 4−5| 5: ≥5)**	[Integration of urban construction conditions 2] KX1_JSTJJC2	5(Higher) 4(High) 3(Medium) 2(Low) 1(Lower)
8.	Reclassification	[Integration of urban construction conditions 2]	[Description] Reclassification (numerical range) [Keywords] C1: <V1|C2: V1 − V2|...|Cn: ≥Vn **KX_Reclass (0: 5,4,3|1: 2| 2: 1)**	[Integration of urban construction conditions] KX1_JSTJJC	
9.	Modification operation	[Utilization conditions of shoreline resource] [Integration of urban construction conditions]	[Description] Raster calculation [Keywords] **KX_RasCalculator([R1] − [R2])**	[Integration of port construction conditions 1] KX1_GKJSJC1	

(continued)

Table 9.6 (continued)

Step	Operation instruction	Input	Operation	Output	Description
10.	Modification value	[Integration of port construction conditions 1]	[Description] Condition calculator [Keywords] **KX_RasCalculator(([R1] < 1), 1|[R1])**	[Integration of port construction conditions] KX1_GKJSJC	
11.	Reclassification	[Annual average risk index of storm surge]	[Description] Reclassifification (numerical range) [Keywords] C1: <V1|C2: V1 − V2|...|Cn ≥Vn **KX_Reclass (1: <2|2: 2–3.5| 3: 3.5–7 |4: ≥7)**	[Marine disaster risk] KX1_HYZH	
12.	Modification operation	[Integration of port construction conditions] [Marine disaster risk]	[Description] Condition calculator [Keywords] **KX_RasCalculator((([R1] = 4) or ([R1] = 5)) and ([R2] = 5), 3|([R1] = 5) and ([R2] = 4), 4|[R1])**	[Suitability of port construction 1] KX1_GKSYX1	
13.	Modification operation	[Location advantage degree] [Suitability of port construction 1]	[Description] Condition calculator [Keywords] **KX_RasCalculator(([R1] = 1), 1|([R1] = 2), [R2] − 1|[R2])**	[Suitability of port construction 2] **KX1_GKSYX2**	
14.	Modification value	[Suitability of port construction 2]	[Description] Condition calculator [Keywords] **KX_RasCalculator (([R1] < 1), 1|[R1])**	[Suitability of port construction] **KX1_GKSYX**	

(continued)

Table 9.6 (continued)

Step	Operation instruction	Input	Operation	Output	Description
15.	Make thematic map	[Suitability of port construction]	[**Description**] Make thematic map [**Keywords**] Style	Background style 1	Background style 2	Resolution **KX_Mapping(C1|Boundary|200)**	[Thematic map of port construction suitability] **KX1_GKSYX.emf**	
16.	Insert thematic map	[Thematic map of port construction suitability]	[**Description**] Insert thematic map [**Keywords**] Picture height **KX_InsertPic (12)**	1	
17.	Produce statistical table	[Statistical layer] [Suitability of port construction]	[**Description**] Make statistical analysis [**Keywords**] Statistical layer, Statistical layer, Display field	5, 4, 3, 2, 1	Table serial number **KX_Statistic (Name	5,4,3,2,1)**	2	

$$[\text{Cultural resources gathering area}] = f([\text{Legal cultural protection space}],$$
$$[\text{Potential cultural protection space}]) \tag{9.6}$$

$$[\text{Contact space of cultural protection}] = f([\text{Legal cultural protection space}],$$
$$[\text{Potential cultural protection space}], [\text{Cultural resources gathering area}]) \tag{9.7}$$

2. **Evaluation steps**

The evaluation process is shown in Fig. 9.2.

(1) Collect relevant information according to Table 9.7 and make [Legal cultural protection space].
(2) [Historical traffic route] is reclassified after Euclidean distance to obtain [Historical traffic route R].
(3) The [Potential historical area] is rasterized and the [Historical traffic route R] takes the maximum value to obtain the [Potential cultural protection space].
(4) [Legal cultural protection space] is rasterized and [Potential cultural protection space] is maximized to obtain [Cultural resources gathering area 1].
(5) [Cultural resources gathering area 1] obtains [Cultural resources gathering area] through aggregation operation and extraction of aggregated patches with a scale of more than 400 ha.

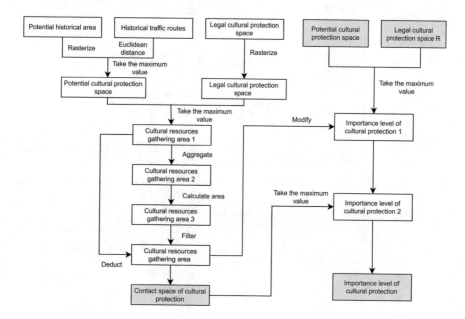

Fig. 9.2 GC flow chart of evaluation of importance of cultural protection

Table 9.7 Reference thresholds of importance level of cultural protection space elements

Major categories of elements	Secondary categories of elements	Protection level	Protecting borders	Importance score
Legal cultural protection space	Immovable historical relics	National level	Scope of protection	5
			Construction control zone	4
		Provincial level	Scope of protection	5
			Construction control zone	4
		City, county (district) level	Scope of protection	4
			Construction control zone	3
		Level not approved	Scope of protection	3
	Historical buildings	–	Core protection scope	3
			Construction control area, style coordination area	2
	World cultural heritage	List	Heritage district	5
			Buffer	4
		Preliminary list	Heritage district	5
			Buffer	4
	Agricultural cultural heritage	World-class	Core protection zone	5
			Overall scope	4
		National level	Core protection zone	5
			Overall scope	4
	Industrial heritage	National level	Core protection zone	5
			Control area	4
		Provincial level	Core protection zone	4
			Control area	3
	Archaeological park	National level	Scope of protection	5
			Construction control zone	4

(continued)

Table 9.7 (continued)

Major categories of elements	Secondary categories of elements	Protection level	Protecting borders	Importance score
			Environmental control area	2
		Provincial level	Scope of protection	4
			Construction control zone	3
			Environmental control area	2
	Famous historical and cultural city	National level	Historical urban area	4
		Provincial level	Historical urban area	3
	Historical and cultural district	–	Core protection scope	5
			Construction control zone	4
			Environmental coordination area	3
	Famous historical and cultural town	National level	Core protection scope	5
			Construction control zone	4
		Provincial level	Core protection scope	4
			Construction control zone	3
	Famous historical and cultural village	National level	Core protection scope	5
			Construction control zone	4
		Provincial level	Core protection scope	4
			Construction control zone	3
	Traditional villages	National level	Scope of protection	5
			Control area	4
		Provincial level	Scope of protection	4
			Control area	3

(continued)

Table 9.7 (continued)

Major categories of elements	Secondary categories of elements	Protection level	Protecting borders	Importance score
	Underground archaeological remains	–	Distribution range	3
	Scenic spots	National level	Scope of core scenic spots	5
			Scope of scenic spots	4
		Provincial level	Scope of core scenic spots	4
			Scope of scenic spots	3
Potential cultural protection space	Potential historical area	–	–	2
	Historical traffic routes	–	–	2
Contact space of cultural protection	–	–	–	1

(6) [Cultural resources gathering area] deducts [Legal cultural protection space] and [Potential cultural protection space] to obtain [Contact space of cultural protection].

(7) [Legal cultural protection space] and [Potential cultural protection space] take the maximum values, and obtain the preliminary evaluation result of [Importance level of cultural protection 1] of [Importance level of cultural protection].

(8) Modify [Importance level of cultural protection 1] with [Cultural resources gathering area]. The importance of [Legal cultural protection space] and [Potential cultural protection space] in the [Cultural resource gathering area] will be raised to one level.

(9) [Importance level of cultural protection 2] and [Contact space of cultural protection] take the maximum values to obtain [Importance level of cultural protection].

Table 9.8 Cultural protection importance [Input and control table]

Serial No.	Layer name	Physical layer	Reference page	Value and description
1.	[Legal cultural protection space]	KX2_FDKJ.shp		Polygon layer, FZ-Score
2.	[Historical traffic routes]	KX2_LSJT.shp		Polyline layer
3.	[Potential historical area]	KX2_QZPQ.shp		Polygon layer
[GC control]				

9.2.2 Input and Control

Table 9.8 shows the [Input and control table] of the importance of cultural protection.

9.2.3 GC Process

Table 9.9 shows the [GC process table] of the importance of cultural protection.

9.2.4 Thematic Map and Statistical Table

Figure 9.3 is the thematic map of cultural protection space, and Table 9.10 is a summary of the evaluation results of the importance of cultural protection.

Table 9.9 [GC process table] of cultural protection importance

Step	Operation instruction	Input	Operation	Output	Description
1.	Rasterization	[Legal cultural protection space]	[Description] Rasterization, the operation field is [Importance] [Keywords] Operation field or value l {Grid size} KX_FeatureToRaster (FZ)	[Legal cultural protection space R] KX2_FDKJ	
2.	Rasterization	[Potential historical area]	[Description] Rasterization, the operation field is [Importance] [Keywords] Operation field or value l {Grid size} KX_FeatureToRaster (2)	[Potential historical area R] KX2_QZPQ	
3.	Euclidean distance + Reclassification	[Historical traffic routes]	[Description] Filter + Euclidean distance + Reclassification [Keywords] {Operation field l Filter list} % Reclassification information KX_SelDisReClass (2: <1000l 0: ≥1000)	[Historical traffic routes R] **KX2_LSJT**	
4.	Take the maximum value	[Potential historical area R] [Historical traffic routes R]	[Description] Take the maximum value [Keywords] KX_Max	[Potential cultural protection space] **KX2_QZKJ**	
5.	Take the maximum value	[Legal cultural protection space R] [Potential cultural protection space]	[Description] Take the maximum value [Keywords] KX_Max	[Cultural resources gathering area 1] **KX2_ZYJJQ1**	
6.	Aggregation operation	[Cultural resources gathering area 1]	[Description] Raster aggregation [Keywords] Aggregation distance KX_Aggregate_Ras (20)	[Cultural resources gathering area 2] **KX2_ZYJJQ1**	

(continued)

Table 9.9 (continued)

Step	Operation instruction	Input	Operation	Output	Description
7.	Patch area calculation	[Cultural resources gathering area 2]	[Description] Patch area calculation [Keywords] Unit KX_GetArea_Ras (km)	[Cultural resources gathering area 3] KX2_ZYJJQ3	
8.	Filter	[Cultural resources gathering area 3]	[Description] Raster calculation [Keywords] Algebraic or logical expression KX_RasCalculator([R1] > 0.4), [R1])	[Cultural resources gathering area] KX2_ZYJJQ	
9.	Clip deduction	[Cultural resources gathering area] [Cultural resources gathering area 1]	[Description] Raster deduction [Keywords] KX_RasErase	[Contact space of cultural protection 1] KX2_LXKJ1	
10.	Value modification	[Contact space of cultural protection 1]	[Description] Raster calculation [Keywords] Algebraic or logical expression KX_RasCalculator([R1] > 0), 1)	[Contact space of cultural protection] KX2_LXKJ	
11.	Take the maximum	[Legal cultural protection space R] [Potential cultural protection space]	[Description] Take the maximum value [Keywords] KX_Max	[Importance level of cultural protection 1] KX2_WHZZX1	
12.	Modification operation	[Cultural resources gathering area] [Importance level of cultural protection 1]	[Description] Raster calculation [Keywords] Algebraic or logical expression KX_RasCalculator([R1] > 0), [R2] + 1)	[Importance level of cultural protection 2] KX2_WHZZX2	

(continued)

Table 9.9 (continued)

Step	Operation instruction	Input	Operation	Output	Description
13.	Take the maximum value	[Importance level of cultural protection 2] [Contact space of cultural protection]	**[Description]** Take the maximum value **[Keywords]** **KX_Max**	[Importance level of cultural protection 3] **KX2_WHZZX3**	
14.	Grade conversion	[Importance level of cultural protection 3]	**[Description]** Reclassification (class modification) **[Keywords]** N1: C1, C2, …IN2 :C3, C4, …I… **KX_Reclass (3: 5I2: 4,3,2I1: 1)**	[Importance level of cultural protection] **KX2_WHZZX**	
15.	Make#	[Importance level of cultural protection]	**[Description]** Make thematic map **[Keywords]** Replace list I Background list I Resolution I {Template} I {Drawing range} **KX_Mapping (S1IBoundary, HI1,I200)**	[Thematic map of importance level of cultural protection] **KX2_WHKJEMF**	
16.	Insert@	[Thematic map of importance level of cultural protection]	**[Description]** Insert thematic map **[Keywords]** Picture heightI {Delete content} **KX_InsertPic (12)**	1	
17.	Make statistical table	[Statistical layer] [Importance level of cultural protection]	**[Description]** Make statistical table **[Keywords]** Display field# {Unit parameter} ICounted type listI {Data filtering mode} **KX_Statistic (NameI3,2,1)**	2	

Table 9.10 The summary of evaluation results of cultural protection importance (unit: area, km²; proportion, %)

Region		Very important		Important		Common	
		Area	Ratio	Area	Ratio	Area	Ratio
* * City	B	3.02	24.0	7.36	58.5	1.63	13.0
	D	16.50	15.2	33.43	30.9	45.10	41.6
	E	42.35	9.3	117.31	25.8	161.58	35.5
	F	26.51	7.0	56.24	14.9	126.35	33.6
	H	10.90	5.8	27.40	14.7	50.69	27.1
	J	2.88	1.9	13.56	8.9	25.78	16.9
	K	13.65	3.3	69.46	16.7	104.79	25.1
	Total	115.80	6.8	324.77	19.0	515.92	30.2

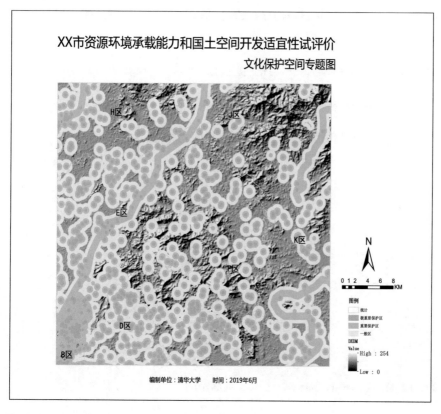

Fig. 9.3 Thematic map of cultural protection space

9.3 Suitability of Mineral Resources Development

9.3.1 Evaluation Method

This model is referred to pages 74–76 in *Technical Guidelines (June Edition)*.

1. **Evaluation methodology**

$$\left[\text{Suitability of mineral resource development and utilization}\right] = f([\text{The value of}$$
$$\text{mineral resource reserves}], \ [\text{Total water}$$
$$\text{resource modulus}], \ [\text{Disaster risk}]) \quad (9.8)$$

$$[\text{The value of mineral resource reserves}] = \text{Sum}([\text{Oil reserves value}],$$
$$[\text{Gas reserves value}], \ \left[\text{Shale gas reserves value}\right], \ldots) \quad (9.9)$$

2. **Evaluation steps**

The evaluation process is shown in Fig. 9.4.

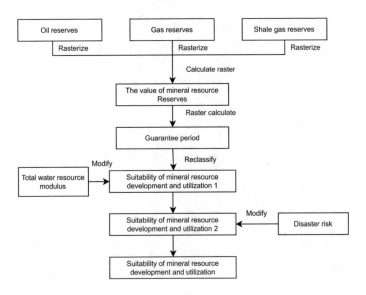

Fig. 9.4 GC flow chart of mineral resources evaluation

Table 9.11 Suitability evaluation of mineral resource development and utilization

Suitability	Not suitable	Less suitable	Generally suitable	More suitable	Suitable
Recommended parameters of security period	<5	5–10	10–20	20–50	≥50

(1) [Oil reserves], [Gas reserves] and [Shale gas reserves] are summed up after rasterization to obtain [The value of mineral resource reserves].

(2) [The value of mineral resource reserves] obtains the [Guarantee period] through the following formula.

$$Y = \frac{f \times t}{P} \tag{9.10}$$

In the formula, Y is the guarantee period, f is the reserve value of mineral resources, t is the recovery rate of industry experience, and P is the output value of the latest year.

(3) The [Guarantee period] is reclassified according to Table 9.11, and the preliminary result [Suitability of mineral resource development and utilization 1] of [Suitability of mineral resource development and utilization] is obtained.

(4) [Suitability of mineral resource development and utilization 1] is modified by using [Total water resource modulus] to obtain [Suitability of mineral resource development and utilization 2].

(5) [Suitability of mineral resource development and utilization 2] is modified by using [Disaster risk] to obtain the final evaluation result of [Suitability of mineral resource development and utilization].

Table 9.12   [Input and control table] of suitability for mineral resource development

Serial No.	Layer name	Physical layer	Reference page	Value and description
1.	[Oil reserves]	KX3_SY.shp		CL-Reserves
2.	[Gas reserves]	KX3_TRQ.shp		CL-Reserves
3.	[Shale gas reserves]	KX3_YYQ.shp		CL-Reserves
4.	[Total water resource modulus]	KX3_SZYZLMS		
5.	[Disaster risk]	KX3_ZHWXX		
	[GC control]			

9.3.2   *Input and Control*

Table 9.12 shows the [Input and control table] of mineral resource development suitability.

9.3.3   *GC Process*

Table 9.13 is the [GC process table] of mineral resource development suitability.

Table 9.13 [GC process table] of mineral resources development suitability

Step	Operation instruction	Input	Operation	Output	Description
1.	Rasterization	[Oil reserves]	**[Description]** Rasterization [**Keywords**] Operation field or value l {Grid size} **KX_FeatureToRaster (CL)**	[Oil reserves R] **KX3_SY**	
2.	Rasterization	[Gas reserves]	**[Description]** Rasterization [**Keywords**] Operation field or value l {Grid size} **KX_FeatureToRaster (CL)**	[Gas reserves R] **KX3_TRQ**	
3.	Rasterization	[Shale gas reserves]	**[Description]** Rasterization [**Keywords**] Operation field or value l {Grid size} **KX_FeatureToRaster (CL)**	[Shale gas reserves R] **KX3_YYQ**	
4.	Summation	[Oil reserves R] [Gas reserves R] [Shale gas reserves R]	**[Description]** Raster calculation [**Keywords**] Algebraic or logical expression **KX_RasCalculator(([R1] + [R2] + [R3])/3)**	[The value of mineral resource reserves] **KX3_KCZYCLJZ**	
5.	Guarantee period calculation	[The value of mineral resource reserves]	**[Description]** Raster calculation [**Keywords**] Algebraic or logical expression **KX_RasCalculator(([R1] * 10/20)**	[Suitability of mineral resource development and utilization 1] KX3_KCZYSYX1	t = 20 P = 20
6.	Reclassification	[Suitability of mineral resource development and utilization 1]	**[Description]** Reclassification (numerical range) [**Keywords**] N1: <V1lN2: V1 − V2l…lNn: ≥Vn **KX_Reclass (1: <5l 2: 5–10l 3: 10–20 l4: 20–50 l 5: ≥50)**	[Suitability of mineral resource development and utilization 2] **KX3_KCZYSYX2**	5(Suitable) 4(More suitable) 3(Generally suitable) 2(Less suitable) 1(Not suitable)

(continued)

Table 9.13 (continued)

Step	Operation instruction	Input	Operation	Output	Description
7.	Modification operation	[Total water resource modulus] [Suitability of mineral resource development and utilization 2]	[Description] Raster calculation If [Total water resource modulus] = 1, then [Suitability of mineral resource development and utilization] − 2 If [Total water resource modulus] = 2, then [Suitability of mineral resource development and utilization] − 1 [Keywords] Algebraic or logical expression KX_RasCalculator(([R1] == 1), [R2] − 2% ([R1] == 2), [R2] − 1%[R2])	[Suitability of mineral resource development and utilization 3] **KX3_KCZYSYX3**	
8.	Modification operation	[Disaster risk] [Suitability of mineral resource development and utilization 3]	[Description] Raster calculation If [Disaster risk] = 4, then [Suitability of mineral resource development and utilization] − 1 If [Disaster risk] = 3, then [Suitability of mineral resource development and utilization] − 1 [Keywords] Algebraic or logical expression KX_RasCalculator(([R1] == 4), [R2] − 1% ([R1] == 3), [R2] − 1%[R2])	[Suitability of mineral resource development and utilization 4] **KX3_KCZYSYX4**	

(continued)

Table 9.13 (continued)

Step	Operation instruction	Input	Operation	Output	Description
9.	Modification value	[Suitability of mineral resource development and utilization 4]	[Description] Raster calculation [Keywords] Algebraic or logical expression **KX_RasCalculator(([R1] < 1), 1 % [R2])**	[Suitability of mineral resource development and utilization 5] **KX3_KCZYSYX5**	
10.	Clip	[Suitability of mineral resource development and utilization 5] [Range layer]	[Description] Clip [Keywords] Clipped layer, clip layer **KX_Clip**	[Suitability of mineral resource development and utilization] **KX3_KCZYSYX**	5(Suitable) 4(More suitable) 3(Generally suitable) 2(Less suitable) 1(Not suitable)
11.	Make thematic map	[Suitability of mineral resource development and utilization]	[Description] Make thematic map [Keywords] Replace list l Background list l Resolution l {Template} l {Drawing range} **KX_Mapping (S1lBoundaryl200)**	[Thematic map of suitability of mineral resource development and utilization] **KX3_KCZYSYX.emf**	
12.	Insert thematic map	[Thematic map of suitability of mineral resource development and utilization]	[Description] Insert thematic map [Keywords] Picture heightl {Delete content} **KX_InsertPic (12)**	1	
13.	Make statistical table	[Statistical layer] [Suitability of mineral resource development and utilization]	[Description] Make statistical table [Keywords] Display field# l {Unit parameter} lCounted type list l {Data filtering mode} **KX_Statistic (Namel5,4,3,2,1)**	2	

Chapter 10
Comparative Analysis

The comparative analysis is mainly based on the evaluation results of the importance of ecological protection, the suitability of agricultural production and the suitability of urban construction and the current situation of land use, the red line of ecological protection, the red line of permanent basic farmland protection, the boundary of urban development and the results of Dual-evaluation at the provincial level. It mainly includes the comparison of the current situation, the comparison of the three-lines, the comparison of province and city, etc.

10.1 Comparison of Current Situation

10.1.1 Method Description

The evaluation results of the importance of ecological protection, the suitability of agricultural production and the suitability of urban construction are overlaid with the data of the Third National Land Survey (hereinafter referred to as the "third land survey data") to find out the existing problems of the current land use and provide support for territorial and spatial planning. The specific analysis and evaluation process are shown in Fig. 10.1.

(1) Extract forest and grassland, water area and water conservancy facilities land, cultivated land, garden plot, urban and village construction land, industrial and mining storage land and transportation land from the third land survey data, and overlay them with ecologically important areas to analyze the distribution of development and utilization land types in ecologically important areas. Refer to Table 10.1 for data extraction.

(2) The paddy field (0101), irrigated land (0102) and dry land (0103) in the third land survey data are extracted and overlaid with the areas with low agricultural

© Surveying and Mapping Press 2021
W. Zhou, *A New GeoComputation Pattern and Its Application in Dual-Evaluation*,
https://doi.org/10.1007/978-981-33-6432-5_10

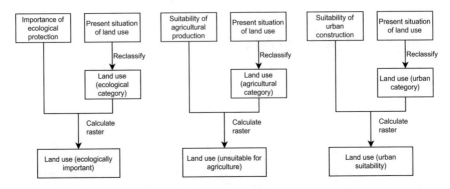

Fig. 10.1 GC flow chart of current situation comparison and evaluation

Table 10.1 Category correspondence table

Analysis classification	Third land survey data classification
Forest and grassland	Forested land (0301, 0302, 0303, 0304), shrubbery (0305, 0306), other woodland (0307), natural grassland (0401, 0402), artificial grassland (0403), other grassland (0404)
Land for water area and water conservancy facilities	River water surface (1101), lake water surface (1102), reservoir water surface (1103), pool water surface (1104), coastal beach (1105), inland beach (1106), ditch (1107, 1108), hydraulic construction land (1109)
Farmland	Paddy field (0101), irrigated land (0102), dry land (0103)
Garden plot	Orchard (0201), tea garden (0202), other gardens (0203, 0204)
Construction land in urban and rural areas	Cities (201), towns (202), villages (203), urban residential land (0701), rural residential land (0702)
Industrial and mining storage land	Industrial land (0601), mining land (0602), storage land (06H1, 201A, 202A, 203A)
Land for transportation	Land for railway (1001), traffic track (1002), highway (1003), rural road (1004), traffic service station (1005), rural road (1006), airport (1007), port and wharf (1008), pipeline transportation (1009)

production suitability, and the distribution of cultivated land in the areas with unsuitable agricultural production is analyzed.

(3) Extract 201 cities, 202 towns and 203 villages from the third land survey data and overlay them with areas with low suitability for urban construction, and analyze the distribution of urban construction land in areas with unsuitable urban construction.

Table 10.2 Status comparison [Input and control table]

Serial No.	Layer name	Physical layer	Reference page	Value and description
1.	[Importance of ecological protection]	JC\JC1\JC1_STZYX		3(Very important)/2(Important)/ 1(Generally important)
2.	[Suitability of agricultural production]	JC\JC2\JC2_NYSYX		3(Suitable)/2(Generally suitable)/1(Not suitable)
3.	[Suitability of urban construction]	JC\JC2\JC3_CZSYX		3(Suitable)/2(Generally suitable)/1(Not suitable)
4.	[Original third land survey data]	DB1_TDLYXZ3D.shp		Polygon layer, DLBM-Land type code
		[GC control]		

10.1.2 Input and Control

Table 10.2 is the [Input and control table] of status comparison.

10.1.3 GC Process

Table 10.3 shows the [GC process table] of current situation comparison.

10.2 Three-lines Comparison

10.2.1 Method Description

The evaluation results of the importance of ecological protection, the suitability of agricultural production and the suitability of urban construction are overlapped with the red line of ecological protection, permanent basic farmland and the boundary of urban development (hereinafter referred to as the three-lines) to judge the existing

Table 10.3 [GC process table] for comparison of current situation

Step	Operation instruction	Input	Operation	Output	Description
1.	Rasterization	[Original third land survey data]	[Description] Rasterization [Keywords] Operation field or value | {Grid size} **KX_FeatureToRaster (DLBM)**	[Third land survey data] DB1_3D	
2.	Extraction of 7 kinds of ground features	[Third land survey data]	[Description] Vector reclassification (class modification) [Keywords] Operation field, Target field% Reclassification information **KX_Reclass_Vec(DLBM|0#1:** **0301,0302,0303,0304,0305,0306,0307,0401,0402,0403,0404|2:** **1101,1102,1103,1104,1105,1106,1107,1108,1109|3:** **0101,0102,0103|4: 0201,0202,0203,0204|5:** **201,202,203,0701,0702|6: 0601,0602,06H1|7:** **1001,1002,1003,1004,1005,1006,1007,1008,1009)**	[Land use (high ecological) 1] DB1_TDST1	1. Forest and grassland 2. Land for water area and water conservancy facilities 3. Farmland 4. Garden plot 5. Construction land in urban and rural areas 6. Industrial and mining storage land 7. Land for transportation
3.	Data filtering	[Importance of ecological protection] [Land use (high ecological) 1]	[Description] Raster calculation If [Importance of ecological protection] = 3, then [Land use (high ecological)] = [Land use (high ecological)1] [Keywords] Algebraic or logical expression **KX_RasCalculator(([R1] == 3), [R2])**	[Land use (high ecological)] ZH3_TDST	

(continued)

Table 10.3 (continued)

Step	Operation instruction	Input	Operation	Output	Description
4.	Make thematic map	[Land use (high ecological)]	**[Description]** Make thematic map **[Keywords]** Replace list l Background list lResolution l {Template} l {Drawing range} **KX_Mapping (TDLYlBoundaryl200)**	[Distribution map of development and utilization land in the most important area of ecological protection] DB1_TDST.EMF	
5.	Insert thematic map	[Distribution map of development and utilization land in the most important area of ecological protection]	**[Description]** Insert thematic map **[Keywords]** Picture heightl {Delete content} **KX_InsertPic (12)**	1	
6.	Make statistical table	[Statistical layer] [Land use (high ecological)]	**[Description]** Make statistical table **[Keywords]** Display field# {Unit parameter} lCounted type list l {Data filtering mode} **KX_Statistic (Namel1,2,3,4,5,6,7)**	2	
7.	Extraction of farmland	[Third land survey data]	**[Description]** Vector reclassification (class modification) **[Keywords]** Operation field, Target field% Reclassification information **KX_Reclass_Vec(DLBMl0#1:0101l2:0102l3:0103)**	[Land use (low agriculture) 1] DB1_TDNY1	1 Paddy field 2 Irrigated land 3 Dry land
8.	Data filtering	[Suitability of agricultural production] [Land use (low agriculture)1]	**[Description]** Raster calculation If [Suitability of agricultural production] = 1, then [Land use (low agriculture)] = [Land use (low agriculture) 1] **[Keywords]** Algebraic or logical expression **KX_RasCalculator(([R1] == 1), [R2])**	[Land use (low agriculture)] DB1_TDNY	

(continued)

Table 10.3 (continued)

Step	Operation instruction	Input	Operation	Output	Description
9.	Make thematic map	[Land use (low agriculture)]	**[Description]** Make thematic map **[Keywords]** Replace list l Background list lResolution l {Template} l {Drawing range} **KX_Mapping (TDlYGDlBoundaryl200)**	[Distribution map of cultivated land in the unsuitable area of agricultural production] DB1_TDST .EMF	
10.	Insert thematic map	[Distribution map of cultivated land in the unsuitable area of agricultural production]	**[Description]** Insert thematic map **[Keywords]** Picture heightl {Delete content} **KX_InsertPic (12)**	3	
11.	Make statistical table	[Statistical layer] [Land use (low agriculture)]	**[Description]** Make statistical table **[Keywords]** Display field# {Unit parameter} lCounted type list l {Data filtering mode} **KX_Statistic (Namel1,2,3)**	4	
12.	Construction landextraction	[Third land survey data]	**[Description]** Vector reclassification (class modification) **[Keywords]** Operation field, Target field% Reclassification information **KX_Reclass_Vec (DLBMl0#1: 201l2:202l3:203)**	[Land use (urban low) 1] DB1_TDCZ1	201. City 202. Town 203. (Village)
13.	Data filtering	[Suitability of urban construction] [Land use (urban low) 1]	**[Description]** Raster calculation If [Suitability of urban construction] = 1, [Land use (urban low)] = [Land use (urban low)1] **[Keywords]** Algebraic or logical expression **KX_RasCalculator(([R1] == 1), [R2])**	[Land use (urban low)] DB1_TDCZ	

(continued)

Table 10.3 (continued)

Step	Operation instruction	Input	Operation	Output	Description
14.	Make thematic map	[Land use (urban low)]	[Description] Make thematic map [Keywords] Replace list ǀ Background list ǀResolution ǀ {Template} ǀ {Drawing range} **KX_Mapping (TDLYCZǀBoundaryǀ200)**	[Distribution map of urban construction land in urban construction unsuitable area] DB1_TDCZ1.EMF	
15.	Insert thematic map	[Distribution map of urban construction land in urban construction unsuitable area]	[Description] Insert thematic map [Keywords] Picture heightǀ {Delete content} **KX_InsertPic (12)**	5	
16.	Make statistical table	[Statistical layer] [Land use (urban low)]	[Description] Make statistical table [Keywords] Display field# {Unit parameter} ǀCounted type list ǀ {Data filtering mode} **KX_Statistic (Nameǀ1,2,3)**	6	

problems in the delineation of the existing three-lines. The specific analysis process is as follows:

(1) Overlay the evaluation results of ecological protection importance with the red line of ecological protection, and analyze the consistency between the red line of ecological protection and the evaluation results of ecological protection importance through thematic maps and statistical tables.
(2) The results of agricultural production suitability evaluation and permanent basic farmland are overlaid, and the consistency between permanent basic farmland and agricultural production suitability evaluation results is analyzed through thematic maps and statistical tables.
(3) The suitability evaluation results of urban construction are overlaid with the urban development boundary, and the consistency between the urban development boundary and the suitability evaluation results of urban construction is analyzed through the form of thematic maps and statistical tables.

10.2.2   Input and Control

Table 10.4 shows the [Input and control table] of the comparison of three control lines.

10.2.3   GC Process

Table 10.5 shows the [GC process table] of the comparison of three-lines.

10.2.4   Thematic Map and Statistical Table

Figure 10.2 is the first thematic map, Fig. 10.3 is the second thematic map, Fig. 10.4 is the third thematic map, Fig. 10.5 is the fourth thematic map. Table 10.6 shows the comparative area and proportion of the importance of ecological protection and the red line of ecological protection, Table 10.7 shows the comparative area and proportion of agricultural production suitability and permanent basic farmland, Table 10.8

Table 10.4 [Input and control table] for the comparison of three-lines

Serial No.	Layer name	Physical layer	Reference page	Value and description
1.	[Ecological red line]	DB2_STHX.shp		Polygon layer
2.	[Permanent basic farmland]	DB2_YJJBNT.shp		Polygon layer
3.	[Current situation of urban land use]	DB2_CZYD_M.shp		Polygon layer
4.	[Current situation of urban and rural land use]	DB2_CXYD_M.shp		Polygon layer
5.	[Importance of ecological protection]	JC\JC1\JC1_STZYX		3(Very important)/2(Important)/ 1(Generally important)
6.	[Suitability of agricultural production]	JC\JC2\JC2_NYSYX		3(Suitable)/2(Generally suitable)/1(Not suitable)
7.	[Suitability of urban construction]	JC\JC3\JC3_CZSYX		3(Suitable)/2(Generally suitable)/1(Not suitable)
8.	[Land use status]	DB\DB1\DB1_TDLYXZ .shp		

[GC control]

shows the comparative area and proportion of urban construction suitability and current urban construction, and Table 10.9 shows the comparative area and proportion of urban construction suitability and current urban and rural construction.

Table 10.5 [GC process table] for the comparison of three control lines

Step	Operation instruction	Input	Operation	Output	Description
1.	Clip	[Importance of ecological protection] [Ecological red line]	**[Description]** Clip **[Keywords]** Clipped layer, Clip layer **KX_Clip**	[Importance of ecological protection 1] DB2_ST1	

(continued)

Table 10.5 (continued)

Step	Operation instruction	Input	Operation	Output	Description
2.	Make thematic map	[Importance of ecological protection] [Ecological red line]	**[Description]** Make thematic map **[Keywords]** Replace list | Background list | Resolution | {Template} | {Drawing range} **KX_Mapping (comparison of importance, ecological red line | boundary | 200)**	[Comparison map between the importance of ecological protection and the red line of ecological protection] DB2_ST2.EMF	
3.	Insert thematic map	[Comparison map between the importance of ecological protection and the red line of ecological protection]	**[Description]** Insert thematic map **[Keywords]** Picture height| {Delete content} **KX_InsertPic (12)**	1	
4.	Make statistical table	[Statistical layer] [Importance of ecological protection 1]	**[Description]** Make statistical table **[Keywords]** Display field# {Unit parameter} |Counted type list | {Data filtering mode} **KX_Statistic (Name|5,4,3,2,1)**	2	
5.	Clip	[Suitability of agricultural production] [Permanent basic farmland]	**[Description]** Clip **[Keywords]** Clipped layer, Clip layer **KX_Clip**	[Suitability of agricultural production 1] DB2_NY1	
6.	Make thematic map	[Suitability of agricultural production]	**[Description]** Make thematic map **[Keywords]** Replace list | Background list | Resolution | {Template} | {Drawing range} **KX_Mapping (Suitability comparison | District boundary, permanent basic farmland|200)**	[Comparison map of agricultural production suitability and permanent basic farmland] DB2_NY2.EMF	
7.	Insert thematic map	[Comparison map of agricultural production suitability and permanent basic farmland]	**[Description]** Insert thematic map **[Keywords]** Picture height| {Delete content} **KX_InsertPic (12)**	3	

(continued)

Table 10.5 (continued)

Step	Operation instruction	Input	Operation	Output	Description
8.	Make statistical table	[Statistical layer] [Suitability of agricultural production 1]	**[Description]** Statistical table **[Keywords]** Display field# {Unit parameter} |Counted type list | {Data filtering mode} **KX_Statistic (Name|5,4,3,2,1)**	4	
9.	Make [Comparison map of urban construction suitability and urban land use status]	[Suitability of urban construction]	**[Description]** Make thematic map **[Keywords]** Replace list | Background list | Resolution | {Template} | {Drawing range} **KX_Mapping (Suitability comparison | District boundary, current situation of urban land|200)**	[Comparison map of urban construction suitability and urban land use status] DB2_CZ1.EMF	
10.	Insert thematic map	[Comparison map of urban construction suitability and urban land use status]	**[Description]** Insert thematic map **[Keywords]** Picture height| {Delete content} **KX_InsertPic (12)**	5	
11.	Make thematic map	[Suitability of urban construction]	**[Description]** Make thematic map **[Keywords]** Replace list | Background list | Resolution | {Template} | {Drawing range} **KX_Mapping (Comparison of suitability | District boundary, current situation of urban and rural land use|200)**	[Comparison map of urban construction suitability and urban and rural land use status] DB2_CZ2.EMF	
12.	Insert thematic map	[Comparison map of urban construction suitability and urban and rural land use status]	**[Description]** Insert thematic map **[Keywords]** Picture height| {Delete content} **KX_InsertPic (12)**	6	
13.	Clip	[Suitability of urban construction] [Current situation of urban land use]	**[Description]** Clip **[Keywords]** Clipped layer, Clip layer **KX_Clip**	[Suitability of urban construction 1] DB2_CZ3	

(continued)

Table 10.5 (continued)

Step	Operation instruction	Input	Operation	Output	Description
14.	Make statistical table	[Statistical layer] [Suitability of urban construction 1]	**[Description]** Make statistical table **[Keywords]** Display field# {Unit parameter} |Counted type list | {Data filtering mode} **KX_Statistic** (Name|5,4,3,2,1)	7	
15.	Clip	[Suitability of urban construction] [Current situation of urban and rural land use]	**[Description]** Clip **[Keywords]** Clipped layer, Clip layer **KX_Clip**	[Suitability of urban construction 2] DB2_CZ4	
16.	Make statistical table	[Statistical layer] [Suitability of urban construction 2]	**[Description]** Make statistical table **[Keywords]** Display field# {Unit parameter} |Counted type list | {Data filtering mode} **KX_Statistic (Name|5,4,3,2,1)**	8	

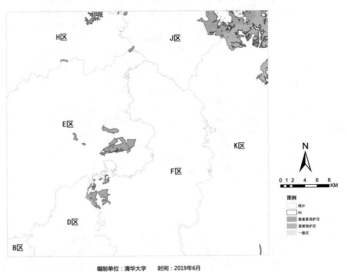

XX市资源环境承载能力和国土空间开发适宜性试评价

生态保护重要性与生态保护红线对比图

编制单位：清华大学   时间：2019年6月

Fig. 10.2 The first thematic map 1

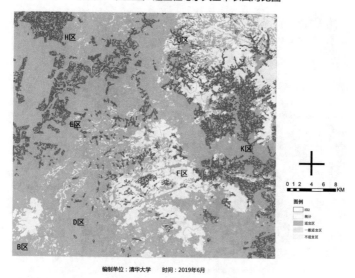

Fig. 10.3   The second thematic map 1

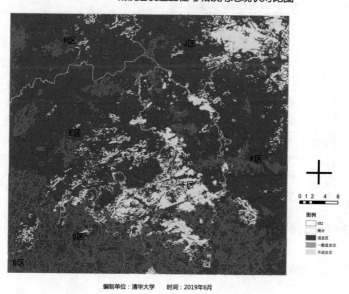

Fig. 10.4   The third thematic map

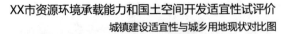

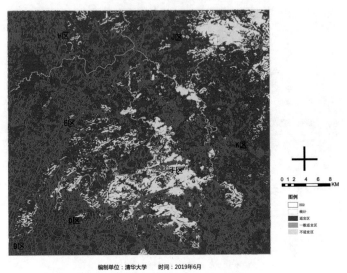

Fig. 10.5 The fourth thematic map

Table 10.6 Comparison area and proportion of ecological protection importance and ecological protection red line (unit: area, km^2; proportion, %)

Region		Excellent		Good		Medium	
		Area	Ratio	Area	Ratio	Area	Ratio
* * City	B	0.00	0.0	0.00	0.0	0.00	0.0
	D	0.00	0.0	0.00	0.0	0.00	0.0
	E	9.41	2.1	0.00	0.0	0.00	0.0
	F	6.35	1.7	0.00	0.0	0.00	0.0
	H	2.17	1.2	0.00	0.0	0.00	0.0
	J	16.95	11.1	0.00	0.0	0.00	0.0
	K	0.28	0.1	24.62	5.9	0.00	0.0
	Total	35.16	2.1	24.62	1.4	0.01	0.0

10.3   Province and City Comparison

10.3.1   Method Description

This GC task is used to compare the evaluation results at the corresponding level with the evaluation results at the higher level, such as the comparison of the evaluation results at the city and county levels with the evaluation results at the provincial level,

Table 10.7 Comparison area and proportion of agricultural production suitability and permanent basic farmland (unit: area, km²; proportion, %)

Region		Suitable		Generally suitable		Not suitable	
		Area	Ratio	Area	Ratio	Area	Ratio
	B	0.00	0.0	0.00	0.0	0.00	0.0
	D	0.63	0.6	0.00	0.0	0.00	0.0
*	E	39.24	8.6	0.00	0.0	0.18	0.0
*	F	15.78	4.2	0.00	0.0	1.98	0.5
City	H	33.49	17.9	0.00	0.0	0.00	0.0
	J	7.37	4.8	0.00	0.0	0.54	0.4
	K	65.42	15.7	0.00	0.0	2.29	0.5
	Total	161.93	9.5	0.00	0.0	4.98	0.3

Table 10.8 Comparison area and proportion of urban construction suitability and current urban construction (unit: area, km²; proportion, %)

Region		Suitable		Generally suitable		Not suitable	
		Area	Ratio	Area	Ratio	Area	Ratio
	B	11.66	92.7	0.00	0.0	0.08	0.6
	D	68.48	63.2	0.00	0.0	0.02	0.0
*	E	64.09	14.1	0.00	0.0	0.13	0.0
*	F	72.68	19.3	0.00	0.0	1.92	0.5
City	H	11.17	6.0	0.00	0.0	0.02	0.0
	J	4.79	3.1	0.00	0.0	0.00	0.0
	K	55.61	13.3	0.00	0.0	0.25	0.1
	Total	288.47	16.9	0.00	0.0	2.41	0.1

Table 10.9 Comparison area and proportion of urban construction suitability and current urban and rural construction (unit: area, km²; proportion, %)

Region		Suitable		Generally suitable		Not suitable	
		Area	Ratio	Area	Ratio	Area	Ratio
	B	11.66	92.7	0.00	0.0	0.08	0.6
	D	68.49	63.2	0.00	0.0	0.02	0.0
*	E	130.25	28.6	0.00	0.0	0.51	0.1
*	F	91.38	24.3	0.00	0.0	3.07	0.8
City	H	36.55	19.6	0.00	0.0	0.06	0.0
	J	16.15	10.6	0.00	0.0	0.04	0.0
	K	70.45	16.9	0.00	0.0	0.30	0.1
	Total	424.92	24.9	0.01	0.0	4.08	0.2

or the comparison of the evaluation results at the provincial level with the evaluation results at the national level to find the difference between the two.

(1) Overlay the results of agricultural production suitability evaluation at the same level with those of higher-level agricultural production suitability evaluation, and analyze the differences between the two through thematic maps and statistical tables.
(2) The suitability evaluation results of urban construction at the same level are overlaid on the suitability evaluation results of urban construction at the higher level, and the differences between the two are analyzed in the form of thematic maps and statistical tables.

10.3.2   Input and Control

Table 10.10 shows the [Input and control table] of province and city comparison.

Table 10.10   [Input and control table] of province and city comparison

Serial No.	Layer name	Physical layer	Reference page	Value and description
1.	[Suitability of agricultural production]	JC\JC2\JC2_NYSYX		3(Suitable)/2(Generally suitable)/1(Not suitable)
2.	[Suitability of urban construction]	JC\JC3\JC3_CZSYX		3(Suitable)/2(Generally suitable)/1(Not suitable)
3.	[Suitability of provincial agricultural production]	NYSYX_GZ		3(Suitable)/2(Generally suitable)/1(Not suitable)
4.	[Suitability of provincial town construction]	CZSYX_GZ		3(Suitable)/2(Generally suitable)/1(Not suitable)

[GC control]

10.3.3 GC Process

Table 10.11 shows the [GC process table] of province and city comparison.

Table 10.11 [GC process table] of province and city comparison

Step	Operation instruction	Input	Operation	Output	Description
1.	Make thematic map	[Suitability of agricultural production]	**[Description]** Make thematic map **[Keywords]** Replace list I Background list IResolution I {Template} I {Drawing range} **KX_Mapping(C2IBoundaryI200)**	[Suitability evaluation chart of agricultural production at city and county level] DB3_ SXDB_3.EMF	
2.	Make thematic map	[Suitability of provincial agricultural production]	**[Description]** Make thematic map **[Keywords]** Replace list I Background list IResolution I {Template} I {Drawing range} **KX_Mapping(C2IBoundaryI200)**	[Suitability evaluation chart of agricultural production at city and county level] DB3_ SXDB _4.EMF	
3.	Insert thematic map	[Suitability evaluation chart of agricultural production at city and county level] [Suitability evaluation chart of agricultural production at city and county level]	**[Description]** Insert thematic map **[Keywords]** Picture heightI {Delete content} **KX_InsertPic (11)**	1	
4.	Statistical analysis (area)	[Statistical layer] [Suitability of agricultural production] [Suitability of provincial agricultural production]	**[Description]** Make statistical table **[Keywords]** Display field# {Unit parameter} ICounted type list I {Data filtering mode} **KX_Statistic (NameI3,2,1I0,1,7,3,9,5,11)**	2	

(continued)

Table 10.11 (continued)

Step	Operation instruction	Input	Operation	Output	Description
5.	Statistical analysis (proportion)	[Statistical layer] [Suitability of agricultural production] [Suitability of provincial agricultural production]	**[Description]** Make statistical table **[Keywords]** Display field# {Unit parameter}	Counted type list	{Data filtering mode} **KX_Statistic (Name|3,2,1|0,2,8,4,10,6,12)**	3	
6.	Make thematic map	[Suitability of urban construction]	**[Description]** Make thematic map **[Keywords]** Replace list	Background list |Resolution	{Template}	{Drawing range} **KX_Mapping(C3|Boundary|200)**	[Suitability evaluation chart of city and county level town construction] DB3_ SXDB_5.EMF	
7.	Make thematic map	[Suitability of provincial town construction]	**[Description]** Make thematic map **[Keywords]** Replace list	Background list |Resolution	{Template}	{Drawing range} **KX_Mapping(C3|Boundary|200)**	[Suitability evaluation chart of provincial urban construction] DB3_ SXDB _6.EMF	
8.	Insert thematic map	[Suitability evaluation chart of city and county level town construction] [Suitability evaluation chart of provincial urban construction]	**[Description]** Insert thematic map **[Keywords]** Picture height| {Delete content} **KX_InsertPic (11)**	4	
9.	Statistical analysis (area)	[Statistical layer] [Suitability of urban construction] [Suitability of provincial town construction]	**[Description]** Make statistical table **[Keywords]** Display field# {Unit parameter}	Counted type list	{Data filtering mode} **KX_Statistic (Name|3,2,1|0,1,7,3,9,5,11)**	5	

(continued)

Table 10.11   (continued)

Step	Operation instruction	Input	Operation	Output	Description
10.	Statistical analysis (proportion)	[Statistical layer] [Suitability of urban construction] [Suitability of provincial town construction]	**[Description]** Make statistical table **[Keywords]** Display field# {Unit parameter} |Counted type list | {Data filtering mode} **KX_Statistic** **(Name|3,2,1|0,2,8,4,10,6,12)**	6	

10.3.4   Thematic Map and Statistical Table

Figure 10.6 is the first thematic map; Fig. 10.7 is the second thematic map. Table 10.12 shows the first comparison between province and city in agricultural production suitability evaluation, Table 10.13 shows the second comparison between province and city in agricultural production suitability evaluation, Table 10.14 shows the first comparison between province and city in urban construction suitability evaluation, and Table 10.15 shows comparison between province and city in urban carrying capacity evaluation.

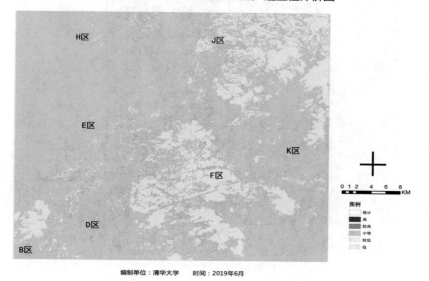

Fig. 10.6 The first thematic map 2

XX市资源环境承载能力和国土空间开发适宜性试评价
省域农业生产适宜性评价图

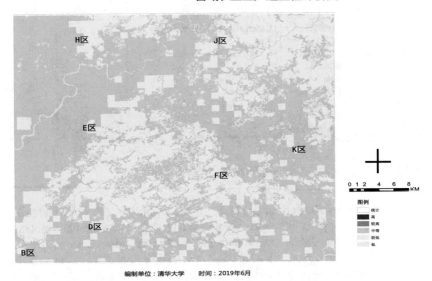

编制单位：清华大学   时间：2019年6月

Fig. 10.6 (continued)

XX市资源环境承载能力和国土空间开发适宜性试评价
市县级城镇建设适宜性评价图

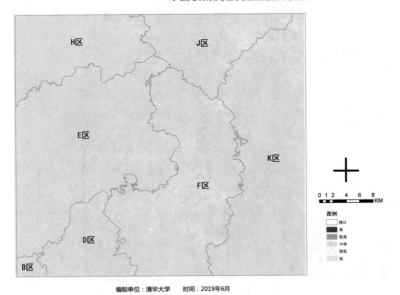

编制单位：清华大学   时间：2019年6月

Fig. 10.7 The second thematic map 2

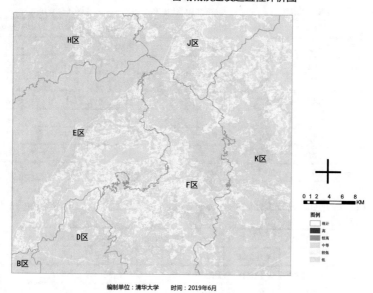

XX市资源环境承载能力和国土空间开发适宜性试评价
省域城镇建设适宜性评价图

编制单位：清华大学   时间：2019年6月

Fig. 10.7 (continued)

Table 10.12 The first comparison between province and city in agricultural production suitability evaluation

Region		Suitable		Generally suitable		Not suitable	
		City	Province	City	Province	City	Province
* * City	B	11.75	9.74	0.00	2.51	0.83	0.33
	D	102.92	74.26	0.00	31.47	5.42	2.61
	E	408.46	274.64	0.00	160.36	46.34	19.79
	F	259.27	210.40	0.00	147.91	116.95	17.91
	H	184.54	141.07	0.00	38.55	2.19	7.11
	J	104.02	58.89	0.00	82.58	48.35	10.90
	K	310.18	240.52	0.00	152.56	107.09	24.19
	Total	1381.14	1009.52	0.00	615.94	327.17	82.83

Table 10.13 The second comparison between province and city in agricultural production suitability evaluation

Region		Suitable		Generally suitable		Not suitable	
		City	Province	City	Province	City	Province
	B	93.4	77.5	0.0	20.0	6.6	2.6
	D	95.0	68.5	0.0	29.1	5.0	2.4
*	E	89.8	60.4	0.0	35.3	10.2	4.4
*	F	68.9	55.9	0.0	39.3	31.1	4.8
City	H	98.8	75.5	0.0	20.6	1.2	3.8
	J	68.2	38.6	0.0	54.1	31.7	7.1
	K	74.4	57.7	0.0	36.6	25.7	5.8
	Total	80.8	59.1	0.0	36.1	19.1	4.8

Table 10.14 Comparison between province and city in urban construction suitability evaluation (unit: area, km^2)

Region		Suitable		Generally suitable		Not suitable	
		City	Province	City	Province	City	Province
	B	12.49	11.50	0.00	0.71	0.09	0.37
	D	104.11	80.62	0.00	11.81	4.23	15.91
*	E	429.65	284.12	0.06	87.37	25.09	83.31
*	F	286.20	203.52	1.13	91.97	88.88	80.72
City	H	182.13	145.11	0.00	21.59	4.59	20.03
	J	123.84	57.63	0.63	42.63	27.91	52.11
	K	376.47	238.61	0.24	86.06	40.57	92.59
	Total	1514.89	1021.11	2.06	342.14	191.35	345.04

Table 10.15 Comparison of province and city for urban bearing capacity evaluation (unit: %)

Region		Suitable		Generally suitable		Not suitable	
		City	Province	City	Province	City	Province
	B	99.3	91.4	0.0	5.7	0.7	2.9
	D	96.1	74.4	0.0	10.9	3.9	14.7
*	E	94.5	62.5	0.0	19.2	5.5	18.3
*	F	76.1	54.1	0.3	24.4	23.6	21.5
City	H	97.5	77.7	0.0	11.6	2.5	10.7
	J	81.2	37.8	0.4	27.9	18.3	34.2
	K	90.3	57.2	0.1	20.6	9.7	22.2
	Total	88.7	59.8	0.1	20.0	11.2	20.2

Chapter 11
Map Analysis

Map analysis is mainly used to evaluate the output of results, including thematic atlas output and questionable point atlas output.

11.1 Thematic Atlas

11.1.1 Method Description

Thematic map atlas is one of the important contents of evaluation results. In order to ensure the consistency between the content of the thematic map atlas and the evaluation computation results, the thematic map atlas needs to be updated according to the results of each computation. This GC task can automatically make thematic maps in batch according to the needs of users and form thematic map atlas in PDF format.

The specific implementation process of this GC task is as follows:

(1) In the completed GC task, select the layer that needs to make thematic maps. In Appendix E.2 of *Technical Guidelines*, there are clear requirements for the maps that need to be output. Of course, users can also select other layers according to their own needs.
(2) For each selected layer, make thematic maps according to the specified display style. It should be noted that there are specific requirements for the display style of specific thematic maps in *Technical Guidelines*.
(3) With KX_Atlas keyword, the thematic maps made are assembled into thematic map atlas. Here, the [Cover] document needs to be prepared in advance.

© Surveying and Mapping Press 2021
W. Zhou, *A New GeoComputation Pattern and Its Application in Dual-Evaluation*,
https://doi.org/10.1007/978-981-33-6432-5_11

11.1.2   Input and Control

Table 11.1 shows the [Input and control table] of thematic atlas.

Table 11.1   Thematic atlas [Input and control table]

Serial No.	Layer name	Physical layer	Reference page	Value and description
1.	[Cover]	Cover1.pdf		
2.	[Topographic features]	DX\DX2\DX2_GC5		
3.	[Administrative division]	BaseMap\MapStat.shp		
4.	[Land use status]	DB\DB1\DB1_TXZK		
5.	[Importance of ecosystem service function]	DX\DX1\DX1_ STZYX		3(Very important)/2(Important)/1(Generally important)
6.	[Ecological sensitivity]	DX\DX1\DX1_ STMGX		3(Extremely sensitive)/2(Sensitive)/1(General sensitive)
7.	[Farming conditions]	DX\DX2\DX2_NYGZTJ		5(Suitable)/4(More suitable)/3(Generally suitable)/2(Less suitable)/1(Not suitable)
8.	[Urban land conditions]	DX\DX2\DX2_CZTDTJ		5(Suitable)/4(More suitable)/3(Generally suitable)/2(Less suitable)/1(Not suitable)
9.	[Agricultural water supply conditions]	DX\DX3\DX3_NYGSTJ		5(Excellent)/4(Good)/3(Common)/2(Poor)/1(Bad)
10.	[Urban water supply conditions]	DX\DX3\DX3_CZGSTJ		5(Excellent)/4(Good)/3(Medium)/2(Poor)/1(Bad)
11.	[Photothermal conditions]	DX\DX4\DX4_GRTJ		5(Excellent)/4(Good)/3(Medium)/2(Poor)/1(Bad)
12.	[Comfort]	DX\DX4\DX4_SSD		5(Excellent)/4(Good)/3(Medium)/2(Poor)/1(Bad)
13.	[Soil environmental capacity]	DX\DX5\DX5_TRHJ		3(Higher)/2(Medium)/1(Lower)

(continued)

Table 11.1 (continued)

Serial No.	Layer name	Physical layer	Reference page	Value and description
14.	[Atmospheric environmental capacity]	DX\DX5\DX5_DQHJ		5(Higher)/4(High)/3(Medium)/ 2(Low)/1(Lower)
15.	[Water environmental capacity]	DX\DX5\DX5_SHJ		5(Higher)/4(High)/3(Medium)/ 2(Low)/1(Lower)
16.	[Geological hazard risk]	DX\DX6\DX6_ZHWXX		4(Higher)/3(High)/2(Medium)/ 1(Lower)
17.	[Location advantage]	DX\DX72\DX72_QWYSD		1(Lower)/2(Low)/3(Medium)/ 4(High)/5(Higher)
18.	[Importance of ecological protection]	JC\JC1\JC1_STZYX		3(Very important)/2(Important)/ 1(Medium)
19.	[Suitability of agricultural production]	JC\JC2\JC2_NYSYX		3(Suitable)/2(Generally suitable)/1(Not suitable)
20.	[Suitability of urban construction]	JC\JC3\JC3_CZSYX		3(Suitable)/2(Generally suitable)/1(Not suitable)
21.	[Agricultural potential space]	ZH\ZH3\ZH3_NYQL		
22.	[Spatial structure of agricultural potential]	ZH\ZH3\ZH3_NYQLJG		
23.	[Urban potential space]	ZH\ZH4\ZH4_CZQL		
24.	[Spatial structure of urban potential]	ZH\ZH4\ZH4_CZQLJG		
25.	[Scenario analysis (Ecology)]	ZH\ZH5\ZH5_QJST		
26.	[Land use (high ecological)]	DB\DB1\ZH3_TDST		
27.	[Land use (low agriculture)]	DB\DB1\ZH3_TDNY		

(continued)

Table 11.1 (continued)

Serial No.	Layer name	Physical layer	Reference page	Value and description
28.	[Land use (urban low)]	DB\DB1\ZH3_TDCZ		
29.	[Ecological red line]	DB\DB2\DB2_STHX.shp		
30.	[Permanent basic farmland]	DB\DB2\DB2_YJJBNT.shp		
31.	[Current situation of urban land use]	DB\DB2\DB2_CZYD_M.shp		
32.	[Current situation of urban and rural land use]	DB\DB2\DB2_CXYD_M.shp		

[GC control]

11.1.3   GC Process

Table 11.2 shows the [GC process table] of the thematic atlas.

11.2   Questionable Point Map

11.2.1   Method Description

In view of the possible questionable points in the evaluation results, the relevant factors affecting the results can be displayed in the form of maps, which is convenient for the evaluators to analyze the problems. The specific implementation process of this function is as follows:

(1) In the completed GC task, select the layer to be displayed.
(2) Specify the display style for each selected layer and make a simplified thematic map.
(3) Show all the simplified thematic maps produced in the designated table.

Table 11.2 [GC process table] of thematic atlas

Step	Operation instruction	Input	Operation	Output	Description
1.	Make thematic map	[Topographic features]	**[Description]** Make thematic map **[Keywords]** Replace list | Background list |Resolution | {Template} | {Drawing range} **KX_Mapping (Topographic features| Boundary |200)**	[Topographic map] DT1_DXDM.pdf	
2.	Make thematic map	[Statistical layer]	**[Description]** Make thematic map **[Keywords]** Replace list | Background list |Resolution | {Template} | {Drawing range} **KX_Mapping (Statistics| Boundary |200)**	[Administrative map] DB1_XZQH.pdf	
3.	Make thematic map	[Land use status]	**[Description]** Make thematic map **[Keywords]** Replace list | Background list |Resolution | {Template} | {Drawing range} **KX_Mapping (TDLY|Boundary|200)**	[Land use status map] DB1_TDLYXZ.pdf	
4.	Make thematic map	[Importance of ecosystem service function]	**[Description]** Make thematic map **[Keywords]** Replace list | Background list |Resolution | {Template} | {Drawing range} **KX_Mapping(S1|Boundary|200)**	[Classification of ecosystem service function importance evaluation results] DX1_STZYX.pdf	

(continued)

Table 11.2 (continued)

Step	Operation instruction	Input	Operation	Output	Description
5.	Make thematic map	[Ecological sensitivity]	[**Description**] Make thematic map [**Keywords**] Replace list | Background list |Resolution | {Template} | {Drawing range} **KX_Mapping(S1|Boundary|200)**	[Classification map of ecological sensitivity evaluation results] DX1_STMGX.pdf	
6.	Make thematic map	[Farming conditions]	[**Description**] Make thematic map [**Keywords**] Replace list | Background list |Resolution | {Template} | {Drawing range} **KX_Mapping(C2|Boundary|200)**	[Grading map of evaluation results of agricultural production land resources] DT1_NYGZTJ.pdf	
7.	Make thematic map	[Urban land conditions]	[**Description**] Make thematic map [**Keywords**] Replace list | Background list |Resolution | {Template} | {Drawing range} **KX_Mapping(C3|Boundary|200)**	[Grading map of urban construction land resources evaluation results] DT1_CZJSTJ.pdf	
8.	Make thematic map	[Agricultural water supply conditions]	[**Description**] Make thematic map [**Keywords**] Replace list | Background list |Resolution | {Template} | {Drawing range} **KX_Mapping(C2|Boundary|200)**	[Classification chart of water resources evaluation results of agricultural production] DT1_NYGSTJ.pdf	
9.	Make thematic map	[Urban water supply conditions]	[**Description**] Make thematic map [**Keywords**] Replace list | Background list |Resolution | {Template} | {Drawing range} **KX_Mapping(C3|Boundary|200)**	[Classification chart of water resources evaluation results of urban construction] DT1_CZGSTJ.pdf	

(continued)

Table 11.2 (continued)

Step	Operation instruction	Input	Operation	Output	Description
10.	Make thematic map	[Photothermal conditions]	**[Description]** Make thematic map **[Keywords]** Replace list I Background list IResolution I {Template} I {Drawing range} **KX_Mapping(C2IBoundaryl200)**	[Classification diagram of Photothermal conditions] DT1_GRTJ.pdf	
11.	Make thematic map	[Comfort]	**[Description]** Make thematic map **[Keywords]** Replace list I Background list IResolution I {Template} I {Drawing range} **KX_Mapping (SSD lBoundaryl200)**	[Comfort index grading chart] DT1_SSD.pdf	
12.	Make thematic map	[Soil environmental capacity]	**[Description]** Make thematic map **[Keywords]** Replace list I Background list IResolution I {Template} I {Drawing range} **KX_Mapping(G3lBoundaryl200)**	[Classification diagram of soil environmental capacity] DX5_TRHJRL.pdf	
13.	Make thematic map	[Atmospheric environmental capacity]	**[Description]** Make thematic map **[Keywords]** Replace list I Background list IResolution I {Template} I {Drawing range} **KX_Mapping(C2IBoundaryl200)**	[Classification diagram of atmospheric environmental capacity] DX5_DQHJK.pdf	
14.	Make thematic map	[Water environmental capacity]	**[Description]** Make thematic map **[Keywords]** Replace list I Background list IResolution I {Template} I {Drawing range} **KX_Mapping(C3IBoundaryl200)**	[Classification diagram of water environment capacity] DX5_SHJK.pdf	

(continued)

Table 11.2 (continued)

Step	Operation instruction	Input	Operation	Output	Description
15.	Make thematic map	[Geological hazard risk]	**[Description]** Make thematic map **[Keywords]** Replace list | Background list |Resolution | {Template} | {Drawing range} **KX_Mapping (ZZZH|Boundary|200)**	[Classification chart of urban construction disaster evaluation results] DX6_ZHWXX.pdf	
16.	Make thematic map	[Location advantage]	**[Description]** Make thematic map **[Keywords]** Replace list | Background list |Resolution | {Template} | {Drawing range} **KX_Mapping(C2|Boundary|200)**	[Location advantage classification map] DX72_ZHYSD.pdf	
17.	Make thematic map	[Importance of ecological protection]	**[Description]** Make thematic map **[Keywords]** Replace list | Background list |Resolution | {Template} | {Drawing range} **KX_Mapping(S1|Boundary|200)**	[Classification chart of importance of ecological protection] DT1_STZYX.pdf	
18.	Make thematic map	[Suitability of agricultural production]	**[Description]** Make thematic map **[Keywords]** Replace list | Background list |Resolution | {Template} | {Drawing range} **KX_Mapping (S2|Boundary|200)**	[Classification chart of agricultural production suitability] DT1_NYSYX.pdf	

(continued)

Table 11.2 (continued)

Step	Operation instruction	Input	Operation	Output	Description
19.	Make atlas	[Cover] [Topographic map] [Administrative map] [Land use status map] [Classification map of sensitivity evaluation results of rocky desertification] [Classification of ecosystem service function importance evaluation results] [Classification map of ecological sensitivity evaluation results] [Grading map of evaluation results of agricultural production land resources] [Grading map of urban construction land resources evaluation results] [Classification chart of water resources evaluation results of agricultural production] [Classification chart of water resources evaluation results of urban construction] [Classification diagram of Photothermal conditions] [Comfort index grading chart] [Classification diagram of soil environmental capacity] ...	**[Description]** Make thematic atlas **[Keywords]** Map list(cover.pdf, Map1.pdf, Map2.pdf, ...) **KX_Atlas**	[Dual-evaluation atlas] DT1_SPJ1004.pdf	

Table 11.3 [Input and control table] of questionable point maps

Serial No.	Layer name	Physical layer	Reference page	Value and description
1.	[questionable point]	SITE2.shp		Area layer
2.	[Importance of ecological protection]	JC\JC1\JC1_STZYX		3(Very important)/2(Important)/ 1(Medium)
3.	[Suitability of agricultural production]	JC\JC2\JC2_NYSYX		3(Suitable)/2(Generally suitable)/1(Not suitable)
4.	[Suitability of urban construction]	JC\JC3\JC3_CZSYX		3(Suitable)/2(Generally suitable)/1(Not suitable)
5.	[Location advantage]	DX\DX72\DX72_Q WYSD		5(Excellent)/4(good)/3(Medium)/ 2(Poor)/1(Bad)
6.	[Urban construction conditions]	DX\DX2\DX2_CZJST J		5(Excellent)/4(good)/3(Medium)/ 2(Poor)/1(Bad)
7.	[Urban water supply conditions]	DX\ DX3\ DX3_CZGSTJ		5(Excellent)/4(good)/3(Medium)/ 2(Poor)/1(Bad)
8.	[Meteorological disaster risk]	DX\ DX6\ DX6_QXZHFX		5(Higher)/4(High)/3(Medium)/2 (Low)/1(Lower)
9.	[Disaster risk]	DX\ DX6\ DX6_ZHWXX		5(Higher)/4(High)/3(Medium)/2 (Low)/1(Lower)
		[GC control]		

11.2.2   Input and Control

Table 11.3 shows the [Input and control table] of questionable point maps.

11.2.3   GC Process

Table 11.4 is the [GC Process table] of questionable point maps.

Table 11.4 [GC process table] of questionable point maps

Steps	Operation instruction	Input	Operation	Output	Description
1.	Make thematic map	[Questionable point] [Importance of ecological protection]	**[Description]** Make thematic map **[Keywords]** Replace list | Background list |Resolution | {Template} | {Drawing range} **KX_Mapping (MARK, C1| 200|2|*)**	[Importance level chart of ecological protection] DT2_DT1.emf	
2.	Make thematic map	[Questionable point] [Suitability of agricultural production]	**[Description]** Make thematic map **[Keywords]** Replace list | Background list |Resolution | {Template} | {Drawing range} **KX_Mapping (MARK, C3| 200|2|*)**	[Suitability map of agricultural production] DT2_DT2.emf	
3.	Make thematic map	[Questionable point] [Suitability of urban construction]	**[Description]** Make thematic map **[Keywords]** Replace list | Background list |Resolution | {Template} | {Drawing range} **KX_Mapping (MARK, C3| 200|2|*)**	[Urban construction suitability map] DT2_DT2.emf	
4.	Make thematic map	[Questionable point] [Location advantage]	**[Description]** Make thematic map **[Keywords]** Replace list | Background list |Resolution | {Template} | {Drawing range} **KX_Mapping (MARK, C2| 200|2|*)**	[Location advantage evaluation chart] DT2_DT3.emf	

(continued)

Table 11.4 (continued)

Steps	Operation instruction	Input	Operation	Output	Description
5.	Make thematic map	[Questionable point] [Urban construction conditions]	[Description] Make thematic map [Keywords] Replace list | Background list |Resolution | {Template} | {Drawing range} **KX_Mapping (MARK, C2| 200|2|*)**	[Evaluation chart of urban construction conditions] DT2_DT4.emf	
6.	Make thematic map	[Questionable point] [Urban water supply conditions]	[Description] Make thematic map [Keywords] Replace list | Background list |Resolution | {Template} | {Drawing range} **KX_Mapping (MARK, C2| 200|2|*)**	[Evaluation chart of urban water supply conditions] DT2_DT5.emf	
7.	Make thematic map	[Questionable point] [Meteorological disaster risk]	[Description] Make thematic map [Keywords] Replace list | Background list |Resolution | {Template} | {Drawing range} **KX_Mapping (MARK, G5| 200|2|*)**	[Meteorological disaster risk evaluation chart] DT2_DT7.emf	

(continued)

Table 11.4 (continued)

Steps	Operation instruction	Input	Operation	Output	Description
8.	Make thematic map	[Questionable point] [Disaster risk]	[**Description**] Make thematic map [**Keywords**] Replace list | Background list |Resolution | {Template} | {Drawing range} **KX_Mapping (MARK, G5| 200|2|*)**	[Hazard evaluation chart] DT2_DT8.emf	
9.	Make atlas	[Importance level chart of ecological protection] [Suitability map of agricultural production] [Urban construction suitability map] [Location advantage evaluation chart] [Evaluation chart of urban construction conditions] [Evaluation chart of urban water supply conditions] [Meteorological disaster risk evaluation chart] [Hazard evaluation chart]	[**Description**] Insert thematic maps [**Keywords**] Picture height| {Delete content} **KX_InsertPic(4|map|)**	1	

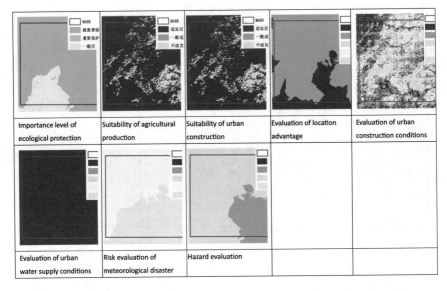

| Importance level of ecological protection | Suitability of agricultural production | Suitability of urban construction | Evaluation of location advantage | Evaluation of urban construction conditions |
| Evaluation of urban water supply conditions | Risk evaluation of meteorological disaster | Hazard evaluation | | |

Fig. 11.1 Questionable point analysis maps

11.2.4   Questionable Point Map

Figure 11.1 is questionable point analysis maps.

Appendix
Layer Style of Thematic Map

Sequence No	Style name	Style ID	Style	RGB value	Value and meaning
1.	Ecological function level	C1	C1 高 较高 中等 较低 低	28,179,2 105,211,89 170,255,190 214,255,213 255,235,190	5 (Higher) 4 (High) 3 (Medium) 2 (Low) 1 (Lower)
2.	Agricultural bearing grade	C2	C2 高 较高 中等 较低 低	109,42,15 231,107,35 247,186,61 255,232,138 214,255,213	5 (Higher) 4 (High) 3 (Medium) 2 (Low) 1 (Lower)
3.	Urban bearing grade	C3	C3 高 较高 中等 较低 低	189,4,38 235,157,147 251,218,213 255,250,194 218,235,193	5 (Higher) 4 (High) 3 (Medium) 2 (Low) 1 (Lower)
4.	Ecological protection level	S1	S1 极重要保护区 重要保护区 一般区	28,179,2 105,211,89 214,255,213	3 (Suitable zone) 2 (Generally suitable zone) 1 (Not suitable zone)

(continued)

© Surveying and Mapping Press 2021
W. Zhou, *A New GeoComputation Pattern and Its Application in Dual-Evaluation*,
https://doi.org/10.1007/978-981-33-6432-5

(continued)

Sequence No	Style name	Style ID	Style	RGB value	Value and meaning
5.	Suitability of agricultural production	S2	☐ S2 ■ 适宜区 ■ 一般适宜区 ■ 不适宜区	250,167, 74 255,224,106 255,254,197	3 (Suitable zone) 2 (Generally suitable zone) 1 (Not suitable zone)
6.	Suitability of urban development	S3	☐ S3 ■ 适宜区 ■ 一般适宜区 ■ 不适宜区	189,4,38 235,157,147 251,218,213	3 (Suitable zone) 2 (Generally suitable zone) 1 (Not suitable zone)
7.	Three kinds of space	3S	☐ 3S ■ 生态空间 ■ 农业空间 ■ 城镇空间	28,179,2 250,167, 74 189,4,38	1 (Ecology) 2 (Agriculture) 3 (Cities and towns)
8.	Environmental factor	HJYZ	☐ HJYZ ■ 好 ■ 较好 ■ 一般 ■ 较差 ■ 差	28,179,2 105,211,89 170,255,190 214,255,213 255,235,190	9 (Excellent) 7 (Good) 5 (Medium) 3 (Poor) 1 (Bad)
9.	Ecological factors	STYZ	☐ STYZ ■ 好 ■ 较好 ■ 一般 ■ 较差 ■ 差	109,42,15 231,107,35 247,186,61 255,232,138 214,255,213	9 (Excellent) 7 (Good) 5 (Medium) 3 (Poor) 1 (Bad)
10.	Ecological space	ST	☐ ST ■ 生态空间	28,179,2	1 (Ecological space)
11.	Agricultural space	NY	☐ NY ■ 农业空间	250,167, 74	1 (Agricultural space)
12.	Urban space	CZ	☐ CZ ■ 城镇空间	189,4,38	1 (Urban space)
13.	Red line	HX	☐ HX ☐	230,0,0	1 (Red line)
14.	Deduction zone	JKQ	☐ JKQ ■ 减扣区	255,0,0	1 (Deduction zone)

(continued)

(continued)

Sequence No	Style name	Style ID	Style	RGB value	Value and meaning
15.	Land type	TDLX	☐ TDLY ■ 林草地 ■ 水域及水利设施 ■ 耕地 ■ 园地 ■ 城镇村建设用地 ■ 工矿仓储用地 ■ 交通运输用地	85,255,0 0,112,255 230,230,0 56,168,0 230,76,0 78,78,78 178,178,178	1 (Forest and grassland) 2 (Land for water area or water conservancy facilities) 3 (Farmland) 4 (Garden plot) 5 (Urban and rural areas construction land) 6 (Industrial and mining storage land) 7 (Land for transportation)
16.	Risk level	G5	☐ G5 ■ 极高 ■ 高 ■ 中 低 ■ 极低	189,4,38 235,157,147 251,218,213 255,250,194 218,235,193	5 (Higher) 4 (High) 3 (Medium) 2 (Low) 1 (Lower)
17.	Questionable point	MARK	☑ MARK ☐	255,0,0	

Printed in the United States
by Baker & Taylor Publisher Services